Structural Safety and Stability of Buildings: Novel Methods, Materials and Measurements

Structural Safety and Stability of Buildings: Novel Methods, Materials and Measurements

Guest Editors

Chaofan Yao
Chuang Zhao
Yan Li

MDPI

Basel • Beijing • Wuhan • Barcelona • Belgrade • Novi Sad • Cluj • Manchester

Guest Editors
Chaofan Yao
School of Civil Engineering
Southwest Jiaotong
University
Chengdu
China

Chuang Zhao
Department of Civil
Engineering and Architecture
Zhejiang University
Hangzhou
China

Yan Li
State Key Laboratory of
Geo-Hazard Prevention and
Geo-Environment Protection
Chengdu University of
Technology
Chengdu
China

Editorial Office
MDPI AG
Grosspeteranlage 5
4052 Basel, Switzerland

This is a reprint of the Special Issue, published open access by the journal *Buildings* (ISSN 2075-5309), freely accessible at: https://www.mdpi.com/journal/buildings/special_issues/33M33KQ60R.

For citation purposes, cite each article independently as indicated on the article page online and as indicated below:

Lastname, A.A.; Lastname, B.B. Article Title. *Journal Name* **Year**, *Volume Number*, Page Range.

ISBN 978-3-7258-4875-1 (Hbk)
ISBN 978-3-7258-4876-8 (PDF)
https://doi.org/10.3390/books978-3-7258-4876-8

Contents

Preface

With the rapid advancement of technology and the ever-evolving nature of society, the construction engineering industry is undergoing unprecedented changes. Traditional construction methods can no longer meet the stringent demands of modern society regarding building quality and efficiency. The industry is now moving towards a more intelligent, green, and sustainable direction.

Research into new methods, materials, and measures is of vital importance to the safety and stability of buildings. Recently developed methods, such as innovative testing techniques or numerical simulation methods, are capable of accurately simulating the stress conditions of buildings under complex loads and considering various nonlinear factors, thereby enhancing the precision of design. New materials, such as advanced composite materials, possess numerous excellent properties that can improve the seismic resistance and fireproof performance of buildings. New measures, such as novel seismic-resistant structural forms, can enhance the resilience of buildings.

This Special Issue encompasses studies on new methods, materials, and measures in various types of structures, including tunnels, pumping stations, transmission towers, and concrete gravity dams. A total of ten papers have been accepted for this Special Issue, all of which are research papers.

Chaofan Yao, Chuang Zhao, and Yan Li
Guest Editors

buildings

MDPI

Article

A Novel Simplified Analysis Model to Predict Behaviors of Single Piles Subjected to Reverse Faulting

Deping Guo [1], Yulin Liu [2], Jincai Tang [1], Zeng Zhang [1], Chaofan Yao [2], Yang Li [2,3,*] and Wang Wu [4,*]

1 Sichuan Shudao Railway Investment Group Co., Ltd., Chengdu 610094, China; guodeping99@163.com (D.G.)
2 State Key Laboratory of Intelligent Geotechnics and Tunnelling, Southwest Jiaotong University, Chengdu 610031, China
3 School of Civil Engineering, Changsha University of Science and Technology, Changsha 410114, China
4 State Key Laboratory of Geo-Hazard Prevention and Geo-Environment Protection, Chengdu University of Technology, Chengdu 610059, China
* Correspondence: liyang@csust.edu.cn (Y.L.); wuwang@cdut.edu.cn (W.W.); Tel.: +86-19848098017 (Y.L.); +86-15520865519 (W.W.)

Abstract: Pile foundations are vulnerable to fault deformations. However, both the physical and numerical modeling of pile foundations under fault deformations are complicated and time-consuming. A simplified model is required for design and engineering practices. This study proposed a novel simplified analysis model to predict the behaviors of single piles subjected to reverse faulting. A two-dimensional beam–spring model is applied. The calculations of the stiffnesses of soil springs, skin friction, ultimate soil resistances, and Young's modulus of sand are presented and discussed. The numerical results show a good agreement with the results of previous centrifuge tests. The parametric studies using the novel model show that ultimate horizontal soil resistance, skin friction, Young's modulus of soil, pile stiffness, and sand density exhibit apparent effects on the responses of a single pile. The ultimate soil resistance controls the maximum inner forces, while Young's modulus affects the increment of inner forces. The bending moment increases with pile stiffness initially and then remains relatively stable. Larger sand density leads to larger inner forces of the pile, owing to greater ultimate soil resistance and stiffness of the soil spring.

Keywords: earthquake; reverse fault; pile; simplified model

1. Introduction

In a previous earthquake, faut-induced damage occurred to structures, such as houses [1,2], bridges [3–6], tunnels [7–11], pipelines [12–14], roads [15], and foundations [16]. The pile foundation was found to be more vulnerable to fault deformation than the shallow foundation. In the 1999 Chi-Chi earthquake, the pile foundations suffered great tilting or sheared off directly by faulting [16]. These instances motivated researchers to explore the responses of pile foundations through testing [17]. Cai and Ng [18] conducted two centrifuge tests for a single pile and a pile group subjected to normal faulting. They found that the single pile was displaced, inducing an additional bending moment along the pile. Li et al. [19] performed sandbox tests for pile groups subjected to thrust faults. They found that the piles in the shear zone could separate. Therefore, the pile foundation should be built far from the shear zone. Yao and Takemura [20] conducted six groups of centrifuge tests for single piles in reverse faulting. Cai et al. [21] summarized two types of loading transfer mechanisms for pile groups subjected to normal faulting using centrifuge tests and numerical studies. Additionally, a series of numerical simulations was applied to investigate the behaviors of piles subjected to faulting [22–25], based on three-dimensional numerical models. As for the anti-measurement, Rasoulia and Fatahib [26] proposed a novel cushioned piled raft foundation to protect buildings from normal faults. Nowkandeh and Ashtiani [27,28] put forward a novel connection between the raft and the piles and a cushioned helical-piled raft system to mitigate hazards in normal faults. Although a series of tests and numerical studies have been conducted for behaviors of piles subjected to faulting, both the physical tests and three-dimensional numerical modeling are complex and time-consuming. It is difficult to handle a large number of models in a short time using physical tests and three-dimensional numerical modeling. Therefore, a simple model is required to assess the pile damage.

In this study, a novel simplified analysis model is proposed to predict the behaviors of single piles subjected to reverse faulting. The novel model is based on a two-dimensional spring model. The input parameters of the novel model are presented and discussed, such as the stiffnesses of the soil springs, skin friction, ultimate soil resistances, and Young's modulus of sand. The objectives of this study include the following: (1) to validate the reasonability of the novel model using previous centrifuge tests; and (2) to discuss the effects of some influencing factors on the pile behaviors in the reverse fault.

2. The Simplified Analysis Model

2.1. General Description of the Novel Model

This study proposed a novel two-dimensional beam–spring model for a single pile subjected to reverse faulting, as illustrated in Figure 1. In the model, the single pile is modeled as a beam, while the surrounding soil is treated as springs. The fault displacements are input through the end of the soil springs, including horizontal displacement Δ_h and vertical displacement Δ_v (Figure 1). Therefore, the interactions between the soil and the pile during faulting are modeled by the deformation of the soil springs. The horizontal springs simulate the lateral resistance from the surrounding soil during faulting. They are set at one side of the pile, which can exhibit both tension and compression. The vertical springs model the skin friction resistance between the pile and surrounding soil. Tip reaction springs for vertical end bearings are not considered as the prototype pile is a floating pile. The pile is meshed to N units with a unit length of ΔL. Therefore, the number of both horizontal and vertical springs is N + 1. The assumptions of the novel model are as follows: (1) the pile is a single pile with an unrestraint pile head; (2) the constitutive model of the pile and the soil springs are bilinear.

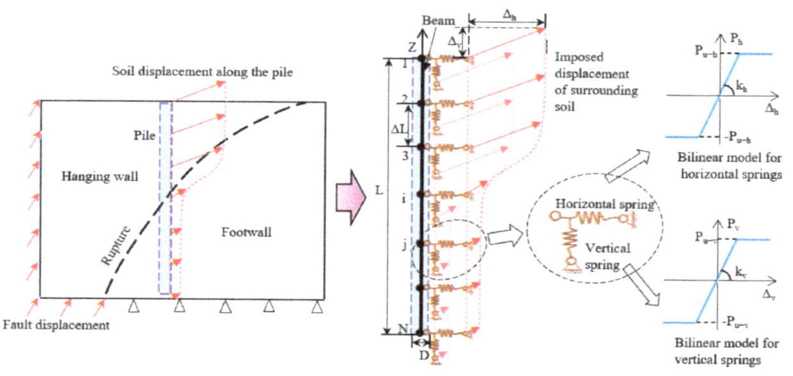

Figure 1. Two-dimensional beam–spring model.

2.2. Stiffnesses and Ultimate Resistances of Horizontal Soil Springs

The horizontal soil springs follow a bilinear constitutive model. Poulos [29] and Pender [30] provided an expression of the bilinear constitutive model for horizontal soil springs to idealize the soil response as follows:

$$\begin{cases} P_h = k_h \Delta_h \\ |P_h| \leq P_{hu} \end{cases}, \tag{1}$$

where P_h, P_{hu}, k_h, and Δ_h are horizontal soil resistance, ultimate horizontal soil resistance, stiffness of horizontal soil spring, and horizontal displacement, respectively.

The stiffness of a horizontal spring has a linear relationship with the coefficient of subgrade reaction, k_s:

$$k_h = k_s D \Delta_L, \tag{2}$$

where D is the diameter of the pile; Δ_L is the unit length of the pile.

According to Equations (1) and (2), the behavior of the horizontal springs is controlled by the coefficient of subgrade reaction, k_s, and the ultimate horizontal soil resistance, P_{hu}. According to previous studies, the coefficient of subgrade reaction, k_s, can be estimated from Young's modulus of soil, E_s, in cohesionless soil, as listed in Table 1. As displayed in Table 1, the coefficient of subgrade reaction ranges from approximately 0.48 to 2 times E_s/D. Therefore, k_s can be evaluated by

$$k_s = C_{ks} E_s / D, \tag{3}$$

where C_{ks} is a constant ranging from 0.48 to 2.

Table 1. Coefficients of subgrade reaction in previous studies.

No.	Author(s)	k_s
1	Terzaghi [31]	$0.74E_s/D$
2	Broms [32]	$(0.48–0.90)E_s/D$
3	Muskhelishvili [33]	$2E_s/D$
4	Matlock [34]	$1.8E_s/D$
5	Poulos [29]	$0.82E_s/D$
6	Rani and Prashant [35]	$0.424{E_s}^{1.136}/D$

For a passive pile in cohesionless soils, Poulos (1995) [29] suggested that the horizontal ultimate resistance could be expressed as

$$P_{hu} = C_{ph} K_P \sigma_v' D \Delta_z, \tag{4}$$

where C_{ph} is a constant ranging from 3 to 5; K_P is the Rankine passive earth pressure coefficient; $K_P = \tan^2(45° + \varphi/2)$, in which φ is the internal frictional angle; $\sigma_{v'}$ is the effective vertical stress; γ_s is the effective unit weight of soil.

Equation (4) was initially proposed by Broms [32] for laterally loaded piles, where C_{ph} was suggested to be 3. However, the value of constant C_{ph} is disputed for taking a value of 3 for passive piles. Some researchers found that the real soil resistance for passive piles was much smaller than $3K_P\,\sigma_{v'}$ when the sand was flowing around the pile, which overestimated the evaluated bending moment of the pile by 2 to 3 times [36]. Therefore, the value of 3 for C_{ph} may also overestimate the bending moment of piles subjected to reverse faulting.

Another uncertainty is that the expression of ultimate horizontal soil resistance may not follow a linear distribution along depth in reverse fault conditions. The faulting leads to a more complicated soil pressure distribution at different locations. Therefore, the linear distribution of ultimate horizontal soil resistance may not satisfy such a complicated soil pressure distribution. The soil on the footwall experiences less compression, which may lead to overestimation of the soil resistance and bending moment of piles on the footwall. To reduce the overestimation, another assumption of ultimate horizontal soil resistance is put forward, as shown in Figure 2. In Assumption 2, the ultimate horizontal soil resistance above the rupture also follows a linear distribution, while it becomes a constant distribution below the rupture. For Assumption 2, this could be expressed as follows:

$$P_{hu} = \begin{cases} C_{ph} K_P \sigma'_v \, D\Delta_z \ (above\ the\ rupture) \\ C_{ph2} P_{hu-r} (below\ the\ rupture) \end{cases}, \tag{5}$$

where C_{ph2} is a constant no larger than 1; P_{hu-r} is the ultimate horizontal soil resistance at the rupture surface.

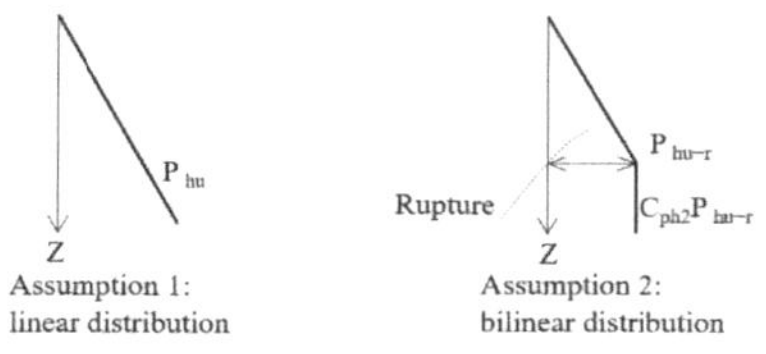

Figure 2. Distribution of ultimate horizontal soil resistance.

2.3. Stiffnesses and Ultimate Resistances of Vertical Soil Springs

The vertical soil springs apply a bilinear constitutive model as well. The expressions are given by

$$\begin{cases} P_v = k_v \Delta_v \\ |P_v| \le P_{vu} \end{cases}, \tag{6}$$

where P_v, P_{vu}, k_v, and Δ_v are the soil resistance, ultimate soil resistance, stiffness, and displacement of the vertical spring, respectively. The stiffness of the soil in the vertical direction can be evaluated according to the work of Randolph and Wroth [37], as follows:

$$k_v = \frac{2\pi G_s \Delta_Z}{In\left(\frac{r_m}{D}\right)}, \tag{7}$$

in which G_s is the shear modulus of the soil, $G_s = E_s/(2(1 + \nu_s))$, $r_m = 2.5L(1 - \nu_s)$, and ν_s is the Poisson's ratio of soil. The ultimate vertical soil resistance is controlled by the sliding friction force on the pile shaft. According to this relationship (Pender [30]), the ultimate skin friction is

$$p_{vu} = C_{pv} K_p \sigma'_v tan\ \varphi_{sp} C_{peri} \Delta_L, \tag{8}$$

where C_{pv} is the constant; φ_{sp} is the interfacial friction angle between the soil and the pile, the suggested values of which are shown in Table 2; C_{peri} is the perimeter of the pile cross-section; $C_{peri} = \pi D$ for a circular pile and $C_{peri} = 4a$ for a square pile; a is the side length of square pile. The distribution of ultimate vertical soil resistance follows Assumption 2 in Figure 2. It could be expressed as follows:

$$P_{vu} = \begin{cases} C_{pv} K_p \sigma'_v tan\ \varphi_{sp} C_{peri} \Delta_L (above\ the\ rupture) \\ C_{pv2} P_{vu-r} (below\ the\ rupture) \end{cases}, \tag{9}$$

where C_{pv2} is a constant no larger than 1, which is suggested to take the same value with C_{ph2}; P_{vu-r} is the ultimate vertical skin friction at the rupture surface, calculated by Equation (8).

Table 2. Suggested friction angle between soil and pile.

Soil Type	Steel Pile	Concrete Pile	Author(s)
Sand	20°	$0.75\varphi_s$	Tomlinson [38]
--	$(0.5–0.7)\varphi_s$	$(0.8–1.0)\varphi_s$	Kulhawy [39]
Dense sand	29.4°	--	O'Neill and Raines [40]
--	$(0.7–0.9)\varphi_s$	--	Yang et al. [41]

2.4. Young's Modulus of Sand

Young's modulus of soil is a key parameter for calculating the stiffness of horizontal springs. There are three Young's modulus distribution models for the static condition (see Figure 3): con (constant soil modulus with depth), par (parabolic variation of soil modulus), and lin (linear variation of soil modulus with depth). For cohesionless soil, it is usually assumed that Young's modulus varies linearly along depth. It is known that a higher stress level leads to a larger Young's modulus. In reverse faulting, the soil above the rupture is highly compressed, while the soil below the rupture is only slightly compressed. A parabolic variation of soil modulus may be more suitable for faulting conditions. Therefore, the Young's modulus applied in the analytical model could be expressed as follows:

$$E_s = \begin{cases} N_h Z\ (Z \leq D) \\ E_{sD} \times \sqrt{Z}/D\ (Z > D) \end{cases}, \quad (10)$$

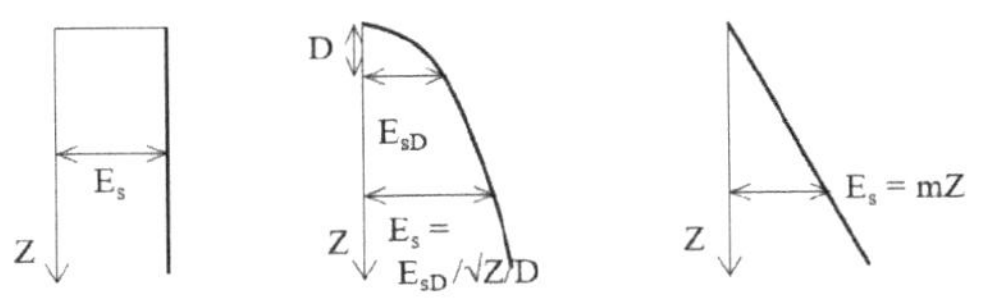

Figure 3. Distributions of Young's modulus with depth.

2.5. Analytical Procedures

The novel method supports a simple approach to evaluate the behaviors of piles subjected to reverse faulting by ANSYS. The analytical procedures are listed as follows:

(1) Prepare the basic parameters of the pile (L, D, EI) and soil (H, φ, ψ, e, γ).
(2) Ensure the number of meshed units N for the pile, and then calculate the stiffnesses of the horizontal and vertical soil springs and their ultimate resistance according to Equations (1), (5), (6), and (8).
(3) Confirm the cross-point location and soil displacements. The location of the cross points and displacements of the surrounding soil could be obtained from the tests or designed values.
(4) Build a beam–spring model by FEM, and calculate the inner forces and displacements of the pile and soil springs.

It should be pointed out that the vertical displacement of the pile and the increase in ground depth on the hanging wall should be considered in Steps 2 and 3.

3. Validation of the Novel Model

Yao and Takemura [20] conducted six groups of centrifuge tests to explore the responses of single piles subjected to faulting, using different fault types, pile locations, and pile stiffness. Three of the tests were conducted under the reverse fault condition. This study applies the square piles of test case RP80 to validate the novel model. The test setups are illustrated in Figure 4, where the prototype scale is presented. The sand with a depth of 11 m was subjected to reverse faulting. The dip angle was $\alpha = 60°$, and the maximum bedrock offset (δb) was $\delta b_{max} = 1.5$ m. Two square piles (SP2 and SP3) were installed at X = 3.75 m and 8.75 m, respectively. The dimensions of the cross-section of the piles were 0.5 m × 0.5 m. The piles had a length of 11 m, where the lower 10 m was buried in the sand. Piles SP2 and SP3 were crossed by the rupture at about Z = −5 m and −2 m, respectively.

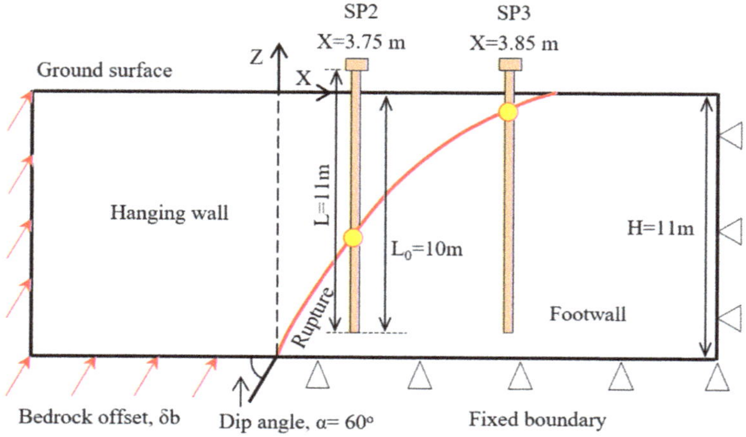

Figure 4. Illustration of the test, adapted from Yao et al. [20].

3.1. Validation by the Square Pile at X = 3.75 m in Case RP80

Table 3 lists the parameters of the pile and sand used in the validation. The pile meshes as 100 units, corresponding to 101 horizontal and vertical springs, respectively. Considering that the frictional angle decreases with shear strain, a 40° frictional angle is applied, which is smaller than the maximum value. C_{ph} is valued as 1 to fit the maximum bending moment after rupture outcrops, which is smaller than that used by Broms [32] and Poulos [29]. C_{pv} is 0.5, meaning that only half of the superficial area of the pile contributes to the effective skin friction. Assumption 2, rather than linear distribution, is applied for the distribution of ultimate horizontal resistance and ultimate skin friction. Both C_{ph2} and C_{pv2} are assigned 0.8. In the validation, the displacements of the soil springs are obtained from the tests.

Table 3. Parameters of pile and sand used in the validation of pile at X = 3.75 m.

Sand	Dr. (%) 80	e 0.682	e_{max} 0.973	e_{min} 0.609	γ (kN/m^3) 15.4
	φ (°) 40	ψ (°) 20	v_s 0.17	N_h (MPa) 5	φ_{sp} 0.5φs
	C_{ks} 1	C_{ph} 1	C_{ph2} 0.8	C_{pv} 0.5	C_{pv2} 0.8
Pile	E_p (Pa) 3.14 × 10^9	L (m) 10	a (m) 0.5	N 100	ΔL (m) 0.1

Figures 5 and 6 compare pile behaviors obtained from the test and calculated from the novel model. In this study, the bending moment is defined as positive when the footwall side of the pile is under tension. The axial force is negative when the pile is under compression. Figure 5a shows that both the evaluated and tested bending moments display S-shaped distributions along the depth, where the extreme values occur at about Z = −5 m and −8 m. It should be pointed out that the input fault displacement of soil may inlcude errors, as it was measured at the front window of the fault simulator rather than at the location of the pile. Therefore, small differences occur in Figure 5a between the tested and estimated bending moments. Figure 5c presents the variations in the extreme values using the bedrock offset. Both the evaluated and tested values in Figure 5c increase initially, and then become relatively stable. The estimated bending moment at Z= −8 m is slightly overestimated from the bedrock offset δb = 0.5 m to 1.1 m. The axial forces distribute as a lying V-shape both for the calculated and tested values (Figure 5b). The extreme values occur at about Z = −6 m, where the rupture crosses the pile. The evaluated axial forces are a little larger than the measured values, experiencing a slightly larger compression. The value at Z = −6 m shows that the calculated axial forces are almost mimicked by the tests (Figure 5d).

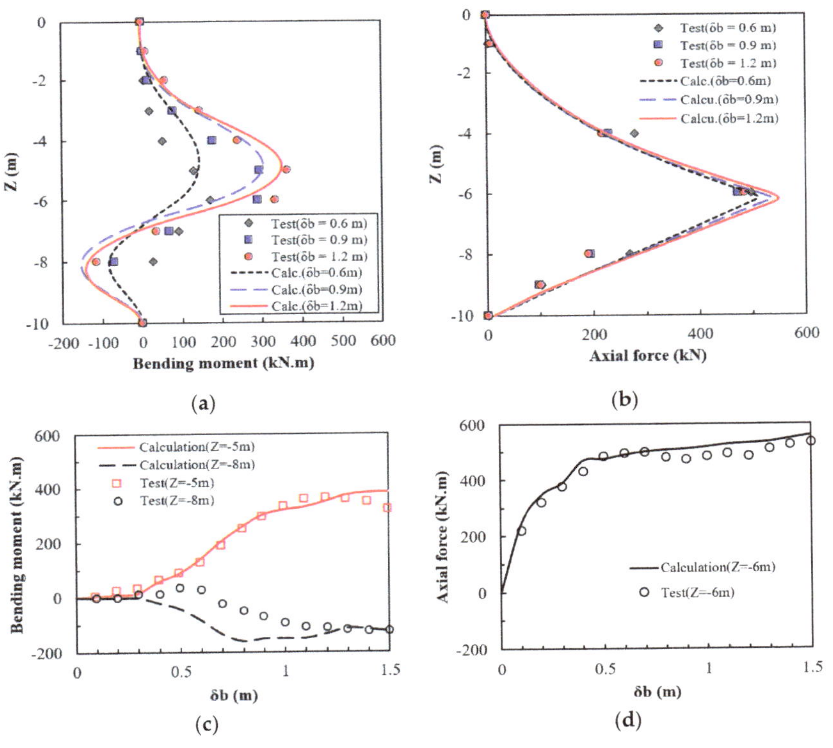

Figure 5. Validation of evaluated inner forces of the pile at X = 3.75 m in Case RP80: (**a**) bending moment distribution; (**b**) axial force distribution; (**c**) bending moment at Z = 5 m and −8 m; (**d**) axial force at Z= −6 m.

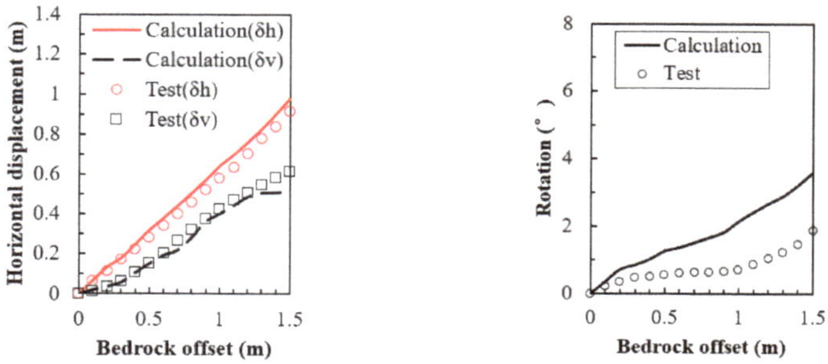

(a) (b)

Figure 6. Evaluated pile top behavior of the pile at X = 3.75 m in Case RP80: (**a**) δh and δv; (**b**) rotation.

Figure 6a shows that both the calculated displacements at the pile top generally agree with those measured by LDT scanning from the test. However, the calculated pile top rotation is larger than that measured from the test (Figure 6b). This may be due to the overestimation of ultimate horizontal soil resistance below the rupture. In general, the novel model yields reasonable results, agreeing with those observed from the test.

3.2. Validation by the Square Pile at X = 8.75 m in Case RP80

Table 4 presents the parameters of the pile at X = 8.75 m with sand applied in the validation. All parameters are the same as those applied for the pile at X = 3.75 m, except C_{ph} and C_{pv}. Their values are 2.7 and 1.35, which are 2.7 times those for the pile at X = 3.75 m. The value of C_{ph} is close to that used by Broms [32] and Poulos [29].

Table 4. Parameters of pile and sand used in the validation of pile at X = 8.75 m.

Sand	Dr (%)	e	e_{max}	e_{min}	γ (kN/m^3)
	80	0.682	0.973	0.609	15.4
	φ (°)	ψ (°)	ν_s	N_h (MPa)	φ_{sp}
	40	20	0.17	5	0.5φ_s
	C_{ks}	C_{ph}	C_{ph2}	C_{pv}	C_{pv2}
	1	2.7	0.8	1.35	0.8
Pile	E_p (Pa)	L (m)	a (m)	N	ΔL (m)
	3.14×10^9	10	0.5	100	0.1

Figure 7 displays the comparison of the evaluated and observed inner forces of the pile. The bending moment presents a D-shaped distribution along the depth (Figure 7a). The estimated bending moments are in better agreement with the tested values above Z = −4 m. The calculated value at Z = −3 m shows a good agreement with the observed measures (Figure 7c). The distributions of axial force are similar for the estimated and measured data, forming a D-shape along the depth (Figure 7b). The axial forces calculated by the novel model are slightly larger than the tested values (Figure 7b). The values at Z = −4 m are generally consistent with each other, especially when δb is larger than 0.5 m.

The comparison of the evaluated and observed displacements and rotation of the pile top are presented in Figure 8. Figure 8a shows that the calculated displacements generally reproduce the values and trends of the tested values. The estimated pile top rotations follow the trend of measured data, with slightly larger values (Figure 8b).

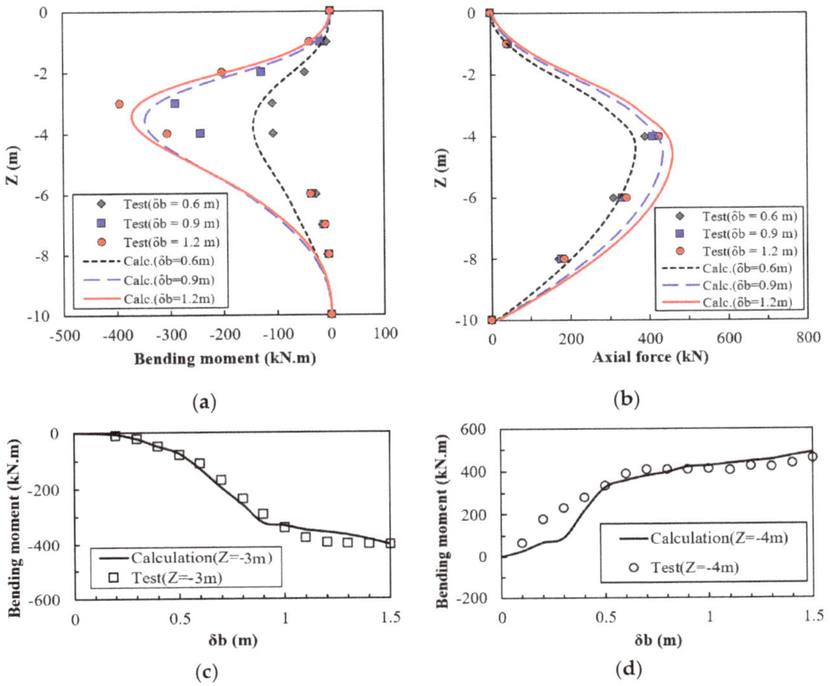

Figure 7. Validation of evaluated inner forces of the pile at X = 8.75 m in Case RP80: (**a**) bending moment distribution; (**b**) axial force distribution; (**c**) bending moment at Z = −3 m; (**d**) axial force at Z= −4 m.

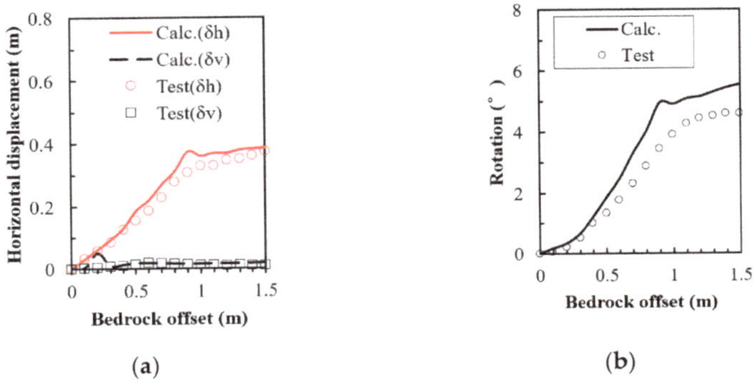

Figure 8. Evaluated top behavior of the pile at X = 8.75 m in Case RP80: (**a**) δh and δv; (**b**) rotation.

Generally, the predictions by the novel model are reasonable, which shows a good agreement with the tested data. The novel model can be applied to the single piles subjected to reverse faulting.

4. Parametric Studies

In this section, the effects of some parameters on the responses of single piles in a reverse fault are discussed, such as ultimate horizontal soil resistance, skin friction, Young's modulus of soil, pile stiffness, and sand density. The parametric studies are based on the tests conducted by Yao and Takemura [20] to evaluate the reasonability of the parameters.

All parameters are the same as those in the validations, unless otherwise stated. The ranges of variables in the parametric studies are listed in Table 5.

Table 5. Ranges of variables in the parametric studies.

Discussed Parameters	Variable	Range of Variable
Ultimate horizontal soil resistance	C_{ph}	0.5–50
skin friction	C_{pv}	0.25–2.5
Young's modulus of soil	N_h	1.5–15
Pile stiffness	EI	9.6–1 $\times$ 10^6 MN·m^2
Sand density	Dr	30–95%

4.1. Ultimate Horizontal Soil Resistance and Skin Friction

The ultimate horizontal soil resistance and skin friction are significant parameters that control the horizontal and vertical responses of the single pile, respectively. In this section, different ultimate horizontal soil resistances and skin frictions are applied for the square piles at X = 3.75 m and 8.75 m in Case RP80, performed by Yao and Takemura [20]. The C_{ph} ranges from 0.5 to 5.0 for the pile at X = 3.75 m, while the C_{pv} is 0.25 to 2.5. For the pile at X = 8.75 m, the ranges of C_{ph} and C_{pv} are 1.0–5.0 and 0.5–2.5, respectively.

Figures 9 and 10 display the evaluated inner forces of single piles with different ultimate soil resistances and skin frictions. As shown in Figure 9a, the bending moment grows with the increase in ultimate soil resistance. The bending moment is limited by the soil stiffness as well, as the values for C_{ks} = 1 become stable after C_{ph} reaches 2.5. The axial force follows a similar trend in which a larger skin friction leads to a larger axial force (Figure 9b,d). The best-fitting parameters for the test data are C_{ph} = 1.0 and C_{pv} = 0.5. For the response of the pile at X = 8.75 m in Figure 10, both the bending moment and the axial force display the same trend as those in Figure 9. The parameters C_{ph} = 3.0 and C_{pv} = 1.35 show the best fit for the experiment. The above results suggest that ultimate soil resistance controls the maximum inner forces. Additionally, it seems that the soil resistances under revere faulting are different at the two locations. Therefore, further discussions are required for the two parameters.

Broms [32] applied 3 $K_p\sigma_v'$ for ultimate horizontal soil resistance in his analysis, which was larger than the passive lateral resistance. Because it corresponds to an infinitely long wall due to lateral stress distribution in the soil, it works well under smaller soil displacement conditions for both active and passive piles. However, the soil displacement is large under reverse faulting, and the sand may flow around the pile. From the results of Yao and Takemura [20], the pile at X = 3.75 m exhibited significantly different horizontal displacements with the soil on the footwall, where the sand flowed around the pile. The one at X = 8.75 m was almost static on the footwall, with very small displacements. Similar phenomena were also observed in previous studies. Viggiani [42] summarized the ultimate horizontal soil resistance values from previous works and applied an ultimate horizontal soil resistance value in the stable soil in his calculation for piles subjected to landslide, which was 2 times that applied for the flowing soil. In Leung et al. [43]'s back analysis for piles subjected to excavation-induced soil movement, they found that 3 $K_p\sigma_v'$ worked well for small soil displacement conditions; however, the evaluated bending moment of the pile was about 2.75 times that of test data when soil displacement was very large flowing around the pile. Smethurst and Powrie [36] also conducted back analysis for piles used to stabilize a railway embankment. They found that horizontal resistance in the flowing sand was about half the limiting pile–soil pressure calculated by 3 $K_p\sigma_v'$. The above findings indicate that the ultimate soil resistances are about 1/3~1/2 of 3 $K_p\sigma_v'$ for passive piles experiencing flowing sand. This agrees with the ultimate soil resistance applied for the pile at X = 3.75 m in this study, which is $K_p\sigma_v{}'$.

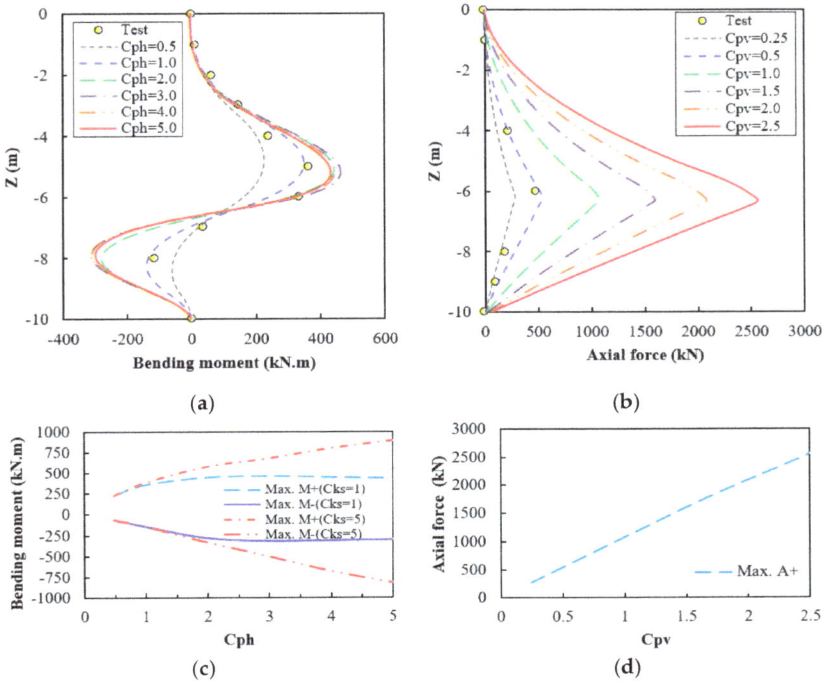

Figure 9. Evaluated inner forces of the pile at X = 3.75 m with different ultimate soil resistances and skin frictions at δb = 1.2 m: (**a**) bending moment distributions under different ultimate soil resistances; (**b**) axical forces distributions under different skin frictions; (**c**) extreme bending moment under different ultimate soil resistances; (**d**) extreme axial force under different skin frictions.

Another major concern for the ultimate soil resistance is the distribution with depth. To further investigate the effects of ultimate soil resistance distribution on pile behaviors, some comparisons of Assumptions 1 and 2 are conducted, based on the novel model. Figure 11 shows the evaluated inner forces of piles with different assumptions of ultimate soil resistance distributions. As presented in Figure 11a, Assumption 1 enlarges the bending moment and axial force of the pile at X = 3.75 m, especially at the lower part of the pile, while Assumption 2 agrees well with the test results. However, the different assumptions have no apparent influence on the inner forces of the pile at X = 8.75 m. This suggests that the pile behaves like an active pile when the rupture crosses the upper part of the pile. This can be calculated using the analytical method for active piles with a linear ultimate soil resistance distribution. However, when the rupture crosses the middle part of the pile, it can no longer be calculated by the analytical method for the active pile. Both the sand flow and a non-linear ultimate soil resistance distribution should be considered.

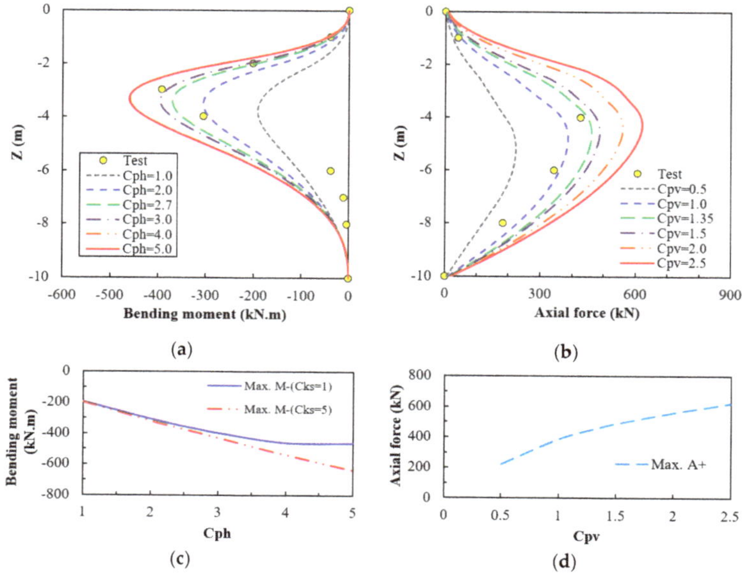

Figure 10. Evaluated inner forces of the pile at X = 8.75 m with different ultimate soil resistances and skin frictions at δb = 1.2 m: (**a**) bending moment distributions under different ultimate soil resistances; (**b**) axial forces distributions under different skin frictions; (**c**) extreme bending moment under different ultimate soil resistances; (**d**) extreme axial force under different skin frictions.

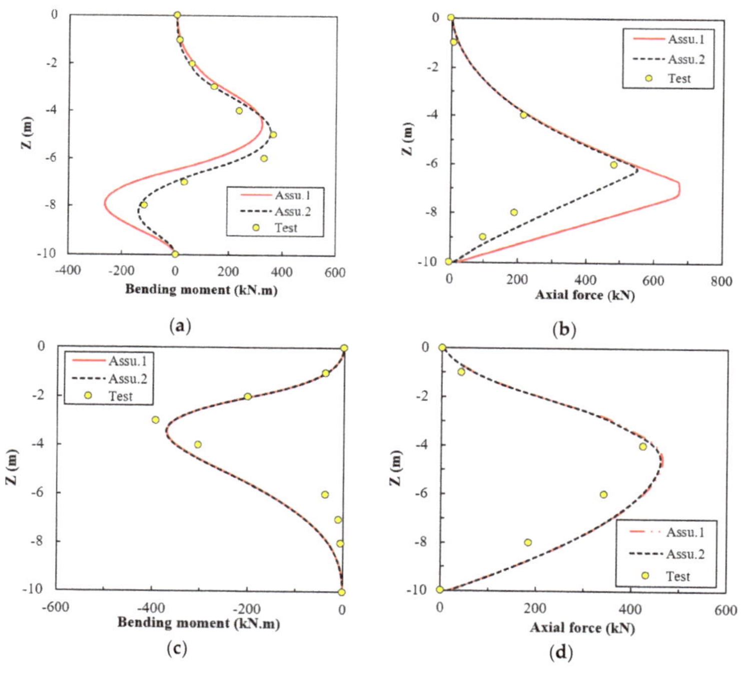

Figure 11. Evaluated inner forces using different assumptions at δb = 1.2 m: (**a**,**b**) bending moment and axial force of the pile at X = 3.75 m, respectively; (**c**,**d**) bending moment and axial force of the pile at X = 8.75 m, respectively.

4.2. Young's Modulus of Soil

Young's modulus of soil is a factor controlling the stiffness of the soil spring. In the novel model, Young's modulus of soil is expressed as N_h times the depth. To investigate the effect of Young's modulus on pile behavior, parametric studies are conducted, where N_h ranges from 1.5 to 15.

Figures 12 and 13 present the evaluated inner forces of piles with different Young's modulus at δb = 1.2 m. As shown in Figure 12, both the bending moment and axial force of the pile at X = 3.75 m increase with Young's modulus and then become relatively stable. This indicates that Young's modulus is a key factor for soil stiffness, affecting the increment of inner forces. However, the inner forces in Figure 12c,d become stable with the limitation of the ultimate soil resistance. In the stable part, the inner forces are controlled by the ultimate soil resistance. For the pile at X = 8.75 m (Figure 13a,c), the bending moment shares the same trend. However, the axial force keeps increasing with Young's modulus (Figure 13b,d). This may be because the lower part of the pile mainly experiences static friction. Therefore, increasing the stiffness of the vertical springs would increase the axial force. In general, a larger Young's modulus leads to larger inner forces before soil resistance reaches the ultimate values.

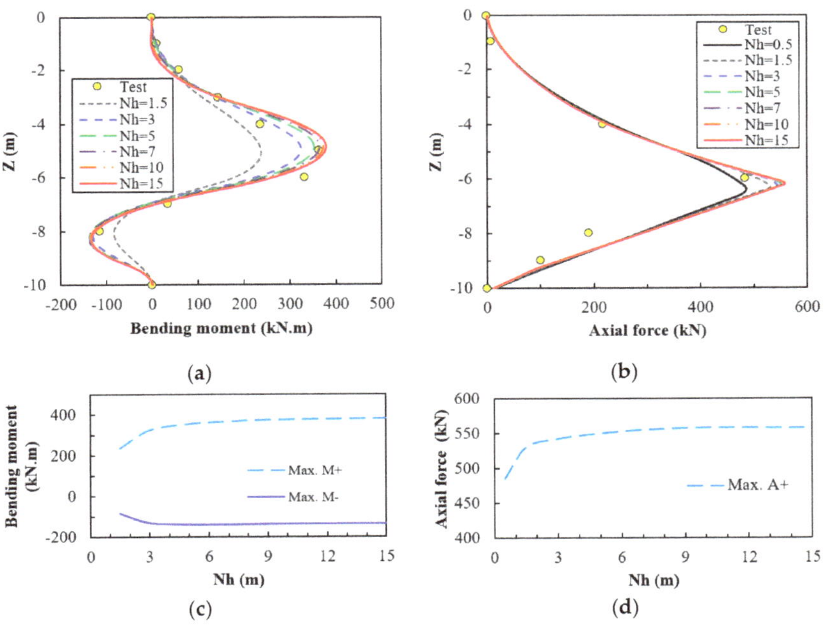

Figure 12. Evaluated inner forces of the pile at X = 3.75 m with different Young's modulus at δb = 1.2 m: (**a**) bending moment distribution; (**b**) axial force distribution; (**c**) extreme bending moment; (**d**) extreme axial force.

4.3. Pile Stiffness

The piles used in the tests of Yao and Takemura [20] are much softer than the actual piles. Figure 14 presents the effects of pile stiffness on pile behaviors. The modeled pile stiffness ranges from 9.6 to 1×10^6 MN·m^2. Additionally, the test data for both the circular piles (CP, EI = 9.6 MN·m^2) and square piles (SP, EI = 16.4 MN·m^2), from Yao and Takemura [20], are displayed.

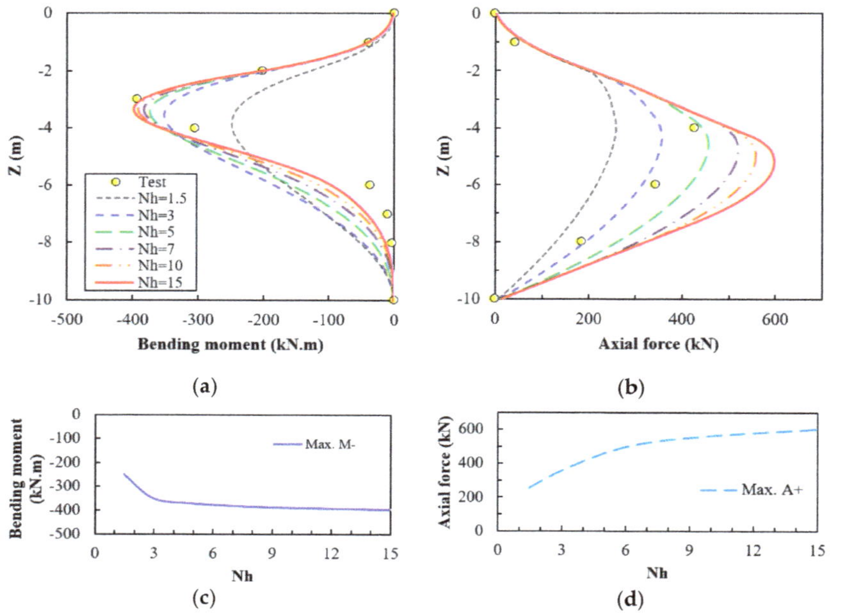

Figure 13. Evaluated inner forces of the pile at X = 8.75 m with different Young's modulus: (**a**) bending moment distribution; (**b**) axial force distribution; (**c**) extreme bending moment; (**d**) extreme axial force.

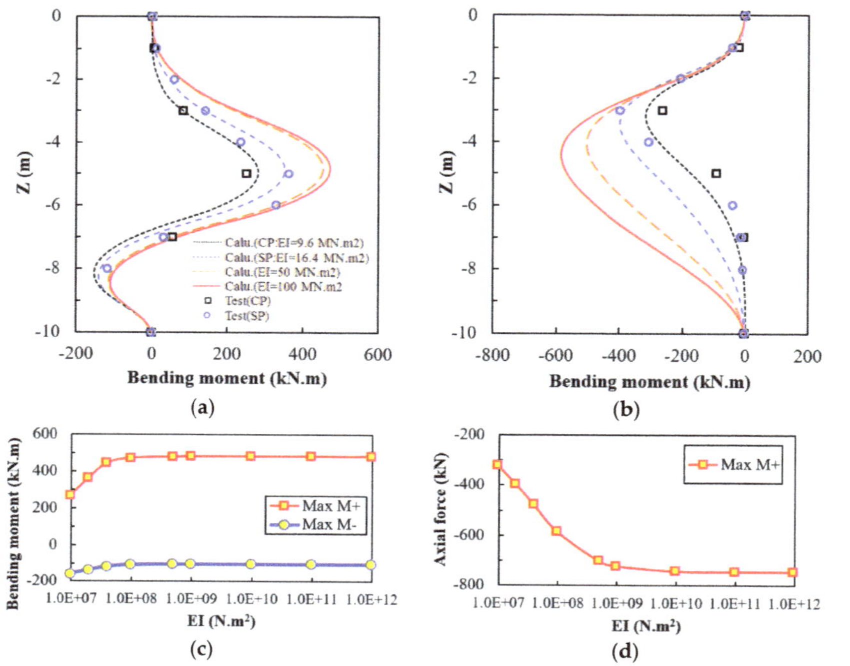

Figure 14. Evaluated inner forces of single piles with different pile stiffness: (**a**,**b**) bending moment distributions of single piles at X = 3.75 m and 8.75, respectively; (**c**,**d**) extreme bending moments at X = 3.75 m and 8.75, respectively.

Figure 14a,b show that the evaluated bending moments of circular and square piles at X = 8.75 m and 13.75 m agree well with the test results. Both bending moment axial forces increase with pile stiffness initially (Figure 14c). However, a larger pile stiffness makes it difficult to induce additional inner forces. The critical point for the bending moment is EI = 100 MN·m^2, which for axial force is EI = 1000 MN·m^2.

4.4. Sand Density

Sand density is a combination of several effect factors, such as void ratio, unit weight, frictional angle, and Young's modulus. In this section, the relative density of sand is considered to be from 30% to 95%, where the test results of Dr = 60% and 80%, obtained by Yao and Takemura [20] are presented as well. The input parameters for the analysis are summarized in Table 6. It should be noted that the calculations ignore the effect of sand density on rupture trace and soil displacement.

Table 6. In put parameters for sand with different densities.

Dr	30	40	50	60	70	80	90	95
e	0.86	0.83	0.79	0.75	0.72	0.68	0.65	0.63
γ_s (kN/m^3)	13.92	14.20	14.49	14.79	15.10	15.43	15.77	15.95
φ_s (°)	31.42	32.82	34.64	36.45	38.43	40	42.14	43.13
N_h (MPa)	1	1.2	1.5	2.5	3.5	5	10	15

Figures 15 and 16 show the evaluated inner forces of piles with different sand densities. As presented in Figure 15, both the bending moment and axial force of the pile at X = 3.75 m increase with sand density. For the pile at X = 8.75 m in Figure 16, the inner forces increase with sand density as well. The increase in inner force is mainly owing to an increment of φ_s and N_h, which enlarge the ultimate soil resistance and the stiffness of the soil springs.

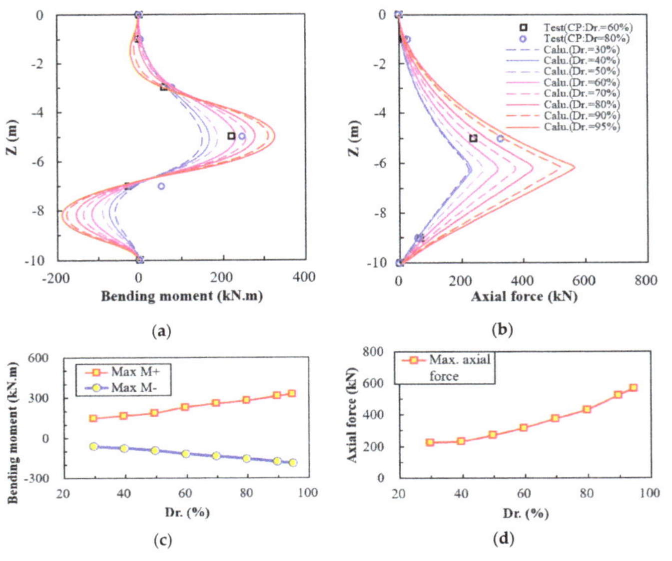

Figure 15. Evaluated inner forces of the pile at X = 3.75 m with sand densities: (**a**) bending moment distribution; (**b**) axial force distribution; (**c**) extreme bending moment; (**d**) extreme axial force.

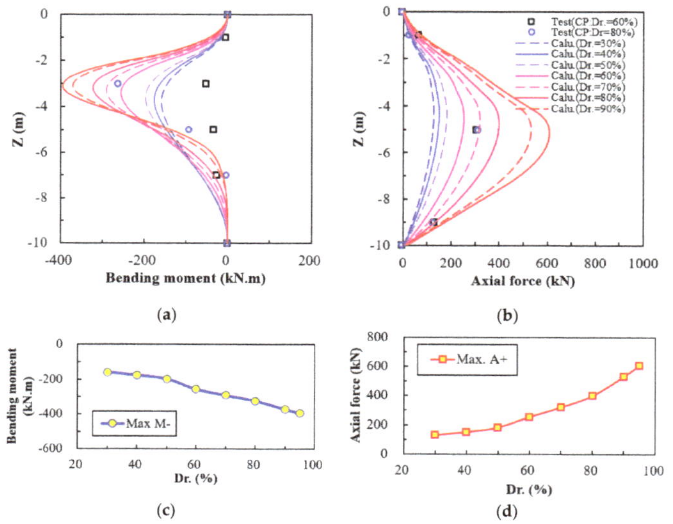

Figure 16. Evaluated inner forces of the pile at X = 8.75 m with sand densities: (**a**) bending moment distribution; (**b**) axial force distribution; (**c**) extreme bending moment; (**d**) extreme axial force.

5. Conclusions

This study proposed a novel simplified analysis model to predict behaviors of single piles subjected to reverse faulting. The novel model was based on a two-dimensional beam–spring model. The calculation methods of the input parameters and the analytical procedures of the novel model are presented. Parametric studies are performed using novel models, such as the ultimate horizontal resistance, skin friction, Young's modulus of soil, pile stiffness, and sand density. The following conclusions are drawn:

(1) The novel model can simulate reasonable results of the inner forces and displacements of single piles. It can be applied to predict behaviors of single piles subjected to reverse faulting.

(2) For passive piles experiencing significant sand flow, an ultimate horizontal soil resistance of $K_p\sigma_v'$ rather than $3K_p\sigma_v'$ is in agreement with the test results. A linear ultimate soil resistance will overestimate the inner forces of the lower part of the pile.

(3) The responses of single piles are affected by the ultimate horizontal soil resistance, skin friction, Young's modulus of soil, pile stiffness, and sand density. The ultimate soil resistance controls the maximum inner forces, while Young's modulus is a key factor for soil stiffness, affecting the increment of inner forces. The bending moment increases with pile stiffness initially, and the increment of pile stiffness rarely induces additional bending moment. Larger sand density leads to larger inner forces of the pile, owing to greater ultimate soil resistances and stiffness of the soil springs.

The novel model is a simplified model, which is easy to operate and is also time-saving. It could be used in design and engineering practices. The application conditions of the novel model are the single piles in sand subjected to reverse faulting. The novel model in this study cannot applied to the group piles, as a three-dimensional model is required. However, the parameters of the soil springs should be modified along the depth according to the geometries when the ground consists of layered sand.

Author Contributions: Methodology, C.Y.; software, Y.L. (Yulin Liu); validation, Y.L. (Yulin Liu) and W.W.; formal analysis, Z.Z.; investigation, J.T.; data curation, Y.L. (Yulin Liu), J.T. and Z.Z.; writing—original draft preparation, D.G.; writing—review and editing, Y.L. (Yang Li) and W.W.; supervision, C.Y.; funding acquisition, Yang Li All authors have read and agreed to the published version of the manuscript.

Funding: This research was funded by Scientific Research Fund of Hunan Provincial Education Department, grant number 24C0141.

Data Availability Statement: The data are contained within the article, and the original contributions presented in this study are also included. Further inquiries can be directed to the corresponding author.

Conflicts of Interest: Authors Deping Guo, Jincai Tang and Zeng Zhang were employed by the company Sichuan Shudao Railway Investment Group Co., Ltd. The remaining authors declare that the research was conducted in the absence of any commercial or financial relationships that could be construed as a potential conflict of interest.

References

1. Anastasopoulos, I.; Gazetas, G. Foundation–Structure Systems over a Rupturing Normal Fault: Part I. Observations after the Kocaeli 1999 Earthquake. *Bull. Earthq. Eng.* **2007**, *5*, 253–275. [CrossRef]
2. Tan, W.; Liang, S.; Ma, X.; Wang, P. Study on the Variation of the Permeability Coefficient of Soil–Rock Mixtures in Fault Zones under Different Stress States. *Deep Undergr. Sci. Eng.* **2024**, 1–12. [CrossRef]
3. Pamuk, A.; Kalkan, E.; Ling, H.I. Structural and Geotechnical Impacts of Surface Rupture on Highway Structures during Recent Earthquakes in Turkey. *Soil Dyn. Earthq. Eng.* **2005**, *25*, 581–589. [CrossRef]
4. Kawashima, K.; Takahashi, Y.; Ge, H.; Wu, Z.; Zhang, J. Reconnaissance Report on Damage of Bridges in 2008 Wenchuan, China, Earthquake. *J. Earthq. Eng.* **2009**, *13*, 965–996. [CrossRef]
5. Yang, S.; Mavroeidis, G.P. Bridges Crossing Fault Rupture Zones: A Review. *Soil Dyn. Earthq. Eng.* **2018**, *113*, 545–571. [CrossRef]
6. Guo, K.; Li, X.; Wang, N.; Wen, Z.; Wang, Y. Effect of Soil–Bridge Interactions on Seismic Response of a Cross-Fault Bridge: A Shaking Table Test Study. *Buildings* **2024**, *14*, 1874. [CrossRef]
7. Yao, C.; He, C.; Wang, T.; Chen, C.; Geng, P.; Dong, W.; Yuan, F.; Xu, G. Damages of Highway Tunnels during 2022 Luding Earthquake (Mw = 6.6). *Soil Dyn. Earthq. Eng.* **2024**, *177*, 108357. [CrossRef]
8. Yao, C.; Luo, W.; Zhang, Y.; He, C.; Li, Y.; Yang, W.; Liu, Y.; Yuan, R. Inclined Rigid Sliding Walls to Protect Tunnels Subjected to Reverse Faulting. *Comput. Geotech.* **2023**, *164*, 105824. [CrossRef]
9. Yao, C.; Luo, W.; Zhang, Y.; He, C.; Yang, W.; Zhang, J.; Yan, Q. A Novel Structural System to Protect Pipe-Jacking Tunnel Subjected to Normal Faulting: Physical Modeling. *Tunn. Undergr. Space Technol.* **2025**, *157*, 106349. [CrossRef]
10. Yu, L.; Zhang, X.; Wang, M.; Yang, H.; Luo, X.; Qin, Y.; Tang, L. A Simplified Analytical Method for the Tunnels' Mechanical Behaviour under the Action of Active Fault Zones. *Comput. Geotech.* **2023**, *164*, 105789. [CrossRef]
11. Cai, Q.P.; Peng, J.M.; Ng, C.W.W.; Shi, J.W.; Chen, X.X. Centrifuge and Numerical Modelling of Tunnel Intersected by Normal Fault Rupture in Sand. *Comput. Geotech.* **2019**, *111*, 137–146. [CrossRef]
12. Yao, C.; He, C.; Huang, X.; Takemura, J.; Yang, W.; An, Z. Response of a Continuous Pipeline in Sand Subjected to Normal Faulting. *Transp. Geotech.* **2022**, *36*, 100824. [CrossRef]
13. Saiyar, M.; Ni, P.; Take, W.A.; Moore, I.D. Response of Pipelines of Differing Flexural Stiffness to Normal Faulting. *Géotechnique* **2016**, *66*, 275–286. [CrossRef]
14. Rojhani, M.; Moradi, M.; Galandarzadeh, A.; Takada, S. Centrifuge Modeling of Buried Continuous Pipelines Subjected to Reverse Faulting. *Can. Geotech. J.* **2012**, *49*, 659–670. [CrossRef]
15. Lu, S.; Chan, A.; Wang, X.; Wang, S.; Wan, Z.; Cheng, J. The Influence of the Disturbing Effect of Roadways through Faults on the Faults' Stability and Slip Characteristics. *Deep Undergr. Sci. Eng.* **2024**, *3*, 399–412. [CrossRef]
16. Chen, C.H.; Chou, H.S.; Yang, C.Y.; Shieh, B.J.; Kao, Y.H. Chelungpu Fault Inflicted Damages of Pile Foundations on FWY Route 3 and Fault Zoning Regulations in Taiwan. In Proceedings of the JSCE/EqTAP Workshop on Seismic Fault-Induced Failures, Tokyo, Japan, 10 February 2003; pp. 1–19.
17. Yao, C.; Takemura, J.; Ma, G.; Dai, C.; An, Z. Effect of Boundary Friction on Revere Fault Rupture Propagation in Centrifuge Tests. *Soil Dyn. Earthq. Eng.* **2021**, *147*, 106811. [CrossRef]
18. Cai, Q.P.; Ng, C.W.W. Centrifuge Modeling of Pile-Sand Interaction Induced by Normal Faulting. *J. Geotech. Geoenviron. Eng.* **2016**, *142*, 04016046. [CrossRef]

19. Li, C.-H.; Lin, M.-L.; Huang, W.-C. Interaction between Pile Groups and Thrust Faults in a Physical Sandbox and Numerical Analysis. *Eng. Geol.* **2019**, *252*, 65–77. [CrossRef]
20. Yao, C.; Takemura, J. Centrifuge Modeling of Single Piles in Sand Subjected to Dip-Slip Faulting. *J. Geotech. Geoenviron. Eng.* **2020**, *146*, 04020001. [CrossRef]
21. Cai, Q.; Xiang, B.; Ng, C.W.W.; Wong, K.S.; Chen, X.; Zhuang, Y. Loading Transfer Mechanism of a Piled Raft Subjected to Normal Faulting in Sand. *Géotech. Lett.* **2022**, *12*, 14–19. [CrossRef]
22. Anastasopoulos, I.; Gazetas, G.; Drosos, V.; Georgarakos, T.; Kourkoulis, R. Design of Bridges against Large Tectonic Deformation. *Earthq. Eng. Eng. Vib.* **2008**, *7*, 345–368. [CrossRef]
23. Anastasopoulos, I.; Kourkoulis, R.; Gazetas, G.; Tsatsis, A. Interaction of Piled Foundation with a Rupturing Normal Fault. *Géotechnique* **2013**, *63*, 1042–1059. [CrossRef]
24. Nooralizadeh Keshteli, O.; Rahimi, S.; Ebadi Jamkhaneh, M. Numerical Investigation of Steel Moment-Resisting Frame on Sandy Soil Under Normal Fault Rupture. *Int. J. Steel Struct.* **2021**, *21*, 703–716. [CrossRef]
25. Alizadeh, M.; Khodaparast, M.; Rajabi, A.M. Simulation of the Interaction of Micropiles and a Fault Rupture. *KSCE J. Civ. Eng.* **2021**, *25*, 4620–4630. [CrossRef]
26. Rasouli, H.; Fatahi, B. A Novel Cushioned Piled Raft Foundation to Protect Buildings Subjected to Normal Fault Rupture. *Comput. Geotech.* **2019**, *106*, 228–248. [CrossRef]
27. Nowkandeh, M.J.; Ashtiani, M. A Novel Mitigation Measure for Normal Fault-Induced Deformations on Pile-Raft Systems. *Earthq. Eng. Eng. Vib.* **2024**, *23*, 15–33. [CrossRef]
28. Jahanshahi Nowkandeh, M.; Ashtiani, M. Cushioned Helical-Piled Raft Systems to Mitigate Hazards Associated with Normal Faulting. *Soil Dyn. Earthq. Eng.* **2023**, *166*, 107773. [CrossRef]
29. Poulos, H.G. Behavior of Laterally Loaded Piles: I-Single Piles. *J. Soil Mech. Found. Div.* **1971**, *97*, 711–731. [CrossRef]
30. Pender, M.J. Aseismic Pile Foundation Design Analysis. *Bull. N. Z. Soc. Earthq. Eng.* **1993**, *26*, 49–160. [CrossRef]
31. Terzaghi, K. Evalution of Conefficients of Subgrade Reaction. *Géotechnique* **1955**, *5*, 297–326. [CrossRef]
32. Broms, B.B. Lateral Resistance of Piles in Cohesionless Soils. *J. Soil Mech. Found. Div.* **1964**, *90*, 123–156. [CrossRef]
33. Muskhelishvili, N.I. Some Basic Problems of the Mathematical Theory of Elasticity: Fundamental Equations, Plane Theory of Elasticity, Torsion, and Bending. *J. Aust. Math. Soc.* **1968**, *8*, b1–b2. [CrossRef]
34. Matlock, H. Correlation for Design of Laterally Loaded Piles in Soft Clay. In Proceedings of the Offshore Technology Conference, Houston, TX, USA, 21–23 April 1970; p. OTC-1204-MS.
35. Rani, S.; Prashant, A. Estimation of the Linear Spring Constant for a Laterally Loaded Monopile Embedded in Nonlinear Soil. *Int. J. Geomech.* **2015**, *15*, 04014090. [CrossRef]
36. Smethurst, J.A.; Powrie, W. Monitoring and Analysis of the Bending Behaviour of Discrete Piles Used to Stabilise a Railway Embankment. *Géotechnique* **2007**, *57*, 663–677. [CrossRef]
37. Randolph, M.F.; Wroth, C.P. Analysis of Deformation of Vertically Loaded Piles. *J. Geotech. Eng. Div.* **1978**, *104*, 1465–1488. [CrossRef]
38. Tomlinson, M.J. *Foundation Design and Construction*; Longman: Harlow, UK, 1986.
39. Kulhawy, F. Limiting Tip and Side Resistance: Fact or Fallacy? *Int. J. Rock Mech. Min. Sci. Geomech. Abstr.* **1986**, *23*, 108. [CrossRef]
40. O'Neill, M.W.; Raines, R.D. Load Transfer for Pipe Piles in Highly Pressured Dense Sand. *J. Geotech. Eng.* **1991**, *117*, 1208–1226. [CrossRef]
41. Yang, J.; Tham, L.G.; Lee, P.K.K.; Chan, S.T.; Yu, F. Behaviour of Jacked and Driven Piles in Sandy Soil. *Géotechnique* **2006**, *56*, 245–259. [CrossRef]
42. Viggiani, C. Ultimate Lateral Load on Piles Used to Stabilize Landslides. In Proceedings of the 10th International Conference on Soil Mechanics and Foundation Engineering (Stockholm), Stockholm, Sweden, 15–19 June 1981; pp. 555–560.
43. Leung, C.F.; Chow, Y.K.; Shen, R.F. Behavior of Pile Subject to Excavation-Induced Soil Movement. *J. Geotech. Geoenviron. Eng.* **2000**, *126*, 947–954. [CrossRef]

buildings

MDPI

Article

Research on the Full Life Cycle Deformation, Stress Response, and Dynamic Fatigue Performance of Concrete Structures in Pump Stations

Min Yuan [1,2,*], Xianhong Zhu [3], Lei Han [3], Yongfeng Qi [1,2], Zhiqiang Xie [1,2], Xi Qin [1,2] and Yun Dong [1,2]

1 Changjiang River Scientific Research Institute, Wuhan 430010, China; qiyf@mail.crsri.cn (Y.Q.); xiezhiqiang@mail.crsri.cn (Z.X.); qinxi@mail.crsri.cn (X.Q.); dongyun@mail.crsri.cn (Y.D.)
2 National Center for Dam Safety Engineering Technology Research, Wuhan 430010, China
3 Anhui Provincial Group Limited for Yangtze-to-Huaihe Water Diversion, Hefei 230000, China; zhuxianhong158@163.com (X.Z.); hanlei@ahyjjh.com.cn (L.H.)
* Correspondence: miny2018@163.com; Tel.: +86-187-0515-0609

Abstract: Concrete structures are subjected to static and fatigue loads during long-term service, and temperature changes can cause changes in the volume of concrete, leading to stress changes and fatigue damage, which affect the mechanical properties and service life of a structure. In response to the current lack of three-dimensional analysis and the exploration of the mechanical behavior and damage evolution processes of concrete structures under variable temperature environmental conditions, this paper relies on the Fenghuangjing Pump Station project involving the Yangtze River and Huaihe River and constructs a concrete thermo-mechanical-fatigue-damage model based on the power exponent function damage constitutive model. By using multiple coupled simulation calculations to predict the performance of concrete structures; analyzing the stress, deformation, and fatigue damage of structures from construction to operation; comparing measured data; dynamically optimizing inversion to obtain concrete thermal parameters; proposing key control parameters; and warning indicators and thresholds, the research results can provide reference for the long-term safe operation of similar projects.

Keywords: pump station concrete; fatigue damage; multi-field coupling; deformation; stress; crack prevention

1. Introduction

Numerous engineering practices have proven that 80% of concrete structure cracks are caused by non-load stresses generated within the concrete material itself [1,2] or under changes in the environmental temperature [3,4] and humidity [5,6]. During long-term service, the pump station structure not only bears static loads [7,8], but also frequently experiences fatigue loads [9,10]. Among them, changes in daily and annual environmental temperatures from the construction period to the operation period can cause volume expansion or contraction in concrete structures, leading to changes in stress and causing fatigue damage. This will cause the internal stress field of a pump station structure to constantly change and be redistributed, leading to the continuous derivation and expansion of microcracks [11,12], resulting in a gradual deterioration of the mechanical properties of structural components [13,14] and increasing the risk of fatigue failure of the structure below the allowable stress [15,16]. The essence of fatigue failure is the process of the continuous accumulation of damage and gradual attenuation of the bearing capacity [17,18].

Numerical simulation, as an effective means of predicting the performance of concrete structures, is particularly important in analyzing the stress [19,20], deformation [21,22], and fatigue damage [23,24] of concrete structures. In terms of coupling algorithms and models, Wang [25] established a relatively new temperature damage model for concrete materials through experiments and calculations. Wang [26] also studied the method of obtaining parameters, describing concrete damage from experiments, and applied temperature damage to the simulation calculation of three-dimensional finite element temperature stress fields. Tang [27] believed that the failure process of concrete actually comprises the initiation and propagation of microcracks until macroscopic cracks are generated, leading to the instability and failure of concrete. A coupled hygro-thermo-mechanical peridynamic model has been proposed for concrete at high temperatures by Zhang [28]. The model can take into account the influences including water vapor diffusion, water vapor flow, and liquid water flow. However, there has not been much analysis of parameters such as the thermal expansion coefficient α, critical fracture energy G_f, and other parameters. The model by Zhang [29] is here employed to investigate the significance of pore pressures and thermally induced stresses for the spalling of concretes with different initial permeabilities and moisture contents through the numerical simulations of a wall and a square column exposed to fire. The numerical results reveal that thermally induced stresses constitute a primary factor in causing damage and hence spalling in concrete while pore pressures, at most, play a secondary role. In order to comprehensively predict the performance of concrete subjected to multiaxial compressive loads in fire, an orthotropic triaxial compression thermo-mechanical damage model for concrete was developed by Zhang [30]. A new numerical simulation framework was introduced by Wang [31] to analyze the long-term damage characteristics of ballastless tracks under cyclic environmental loading considering the thermal properties of the track and the nonlinearity of concrete. In terms of experimentation, the experiments by Seng [32] allowed researchers to demonstrate how the heat and moisture transport phenomena within a wall are coupled, particularly how a temperature difference can be a sufficient driving force for the release of moisture. The author's work points out the impact of moisture adsorption on heat release and on the temperature changes within the wall. The findings by Zhu [33] demonstrated that temperature significantly impacts relative humidity development and that temperature and relative humidity are the primary determinants of early stress development in concrete. The effects of mineral admixtures and curing ages on the hydration process, mechanical properties, and permeability of concrete during the steam curing process were studied by Shi [34], and the pore structure, micromorphology, and interfacial transition zone (ITZ) were also determined by multi-scale image processing and scanning electron microscopy.

Although the above references have conducted extensive researches on the stress, de-formation, and damage of concrete structures, there is still a lack of three-dimensional analysis and the exploration of the mechanical behavior and damage evolution processes of early-age concrete structures in environments with changing temperature and humidity. In particular, the influences of damage on thermal diffusivity, the thermal conductivity coefficient, and elastic modulus still need to be studied. Therefore, the proposal and establishment of multi-field coupling algorithms and models for thermo-mechanical-damage are necessary.

Therefore, in order to comprehensively understand and deal with fatigue damage under multi-field coupling conditions from a three-dimensional perspective, this paper relies on the Fenghuangjing Pump Station project involving the Yangtze River and Huaihe River, establishes a concrete thermo-mechanical-fatigue-damage model based on the power exponent function damage constitutive model [26], and uses the Fortran language to develop a corresponding three-dimensional finite element simulation calculation program. The finite element mesh model of the pump station structure is subjected to thermo-mechanical-fatigue-damage multi-field coupling simulation calculation and the adiabatic temperature rise of the material is inverted to obtain a concrete structure prediction model. On this basis, the numerical simulation prediction and analysis of the entire process of the pump station structure are carried out to obtain performance response parameters such as the stress performance, deformation performance, and dynamic fatigue damage of the pump station structure throughout the entire life cycle from construction to operation. These parameters are compared with the response characteristics (monitoring values) of the physical structure, and key control structural parameters, warning indicators, and thresholds for long-term safe operation of the structure are proposed. Through exploring the above research content, the aim is to provide reference for ensuring the long-term safe and efficient operation of the structure.

2. Project Overview

Located on the north bank of the Yangtze River embankment in Liudu Town, Wuwei City, Anhui Province of China, the Fenghuangjing Pump Station project is the first diversion hub of the Xizhao Canal for the Yangtze-River-to-Huaihe-River Water Diversion Project, and it was also the last but crucial project to commence among the eight major hubs of the "Infrastructure Construction Project No.1" of the Yangtze-River-to-Huaihe-River Water Diversion Project in Anhui Province. The project aims to optimize the allocation of water resources through cross-basin water transfer, alleviate the problem of drought and water shortage in the Huai River Basin, improve the ecological environment of Chaohu Lake and the Huai River, and promote the coordinated development of the Yangtze River Economic Belt and the Huai River Ecological Economic Belt. The Fenghuangjing Pump Station is located south of the Jianghuai watershed, belonging to the Yangtze River Basin. It is in a warm temperate and northern subtropical humid monsoon climate zone with a mild climate, four distinct seasons, abundant sunshine, and long frost-free period. The dominant wind directions in this area are northeast (winter) and southwest (summer), with an average annual wind speed of 2 m/s.

3. Main Calculation Parameters

3.1. Temperature Data

In the simulation calculation, the formula for fitting the average monthly temperature of the concrete structure of the Fenghuangjing Pump Station into a cosine curve over many years is as follows [35] while considering a day–night temperature difference of ±10.0 °C.

$$T_a(t) = 16.6 + 12.3 \times cos\left[\frac{\pi}{6}(t - 7.2)\right] \tag{1}$$

3.2. Main Thermodynamic Parameters

The main thermodynamic parameters of the foundation refer to the engineering data and similar projects [35]. Concrete specimens of the C30 target strength grade were prepared according to the Chinese standard JGJ55-2011 [36] the thermodynamic parameters of concrete (C30) were obtained from indoor tests, and the above parameters are detailed in Table 1.

Table 1. Thermodynamic parameters of the foundation and concrete.

Category	Thermal Conductivity Coefficient λ/ ($kJ{\cdot}m^{-1}{\cdot}h^{-1}{\cdot}{}^{\circ}C^{-1}$)	Specific Heat c/ ($kJ{\cdot}kg^{-1}{\cdot}{}^{\circ}C^{-1}$)	Thermal Diffusivity a/ ($m^2{\cdot}h^{-1}$)	Final Adiabatic Temperature Rise θ_0/ °C	Linear Expansion Coefficient α/ ($10^{-6}{\cdot}{}^{\circ}C^{-1}$)	Final Modulus of Elasticity E_0/ (MPa)	Poisson's Ratio μ
Foundation	10.838	0.886	0.004780	/	6.8	300	0.2
C30	7.830	0.870	0.003524	41.6	8.1	30,000	0.167

$$\text{Elastic modulus of concrete (GPa)}: \; E(\tau) = 39.6 \times \left[1 - e^{-0.39\tau^{0.496}}\right] \tag{2}$$

$$\text{Concrete compressive strength (MPa)}: \; f_c(\tau) = 47.5 \times \left[1 - e^{-0.49\tau^{0.497}}\right] \tag{3}$$

$$\text{Concrete tensile strength (MPa)}: \; f_t(\tau) = 2.96 \times \left[1 - e^{-0.46\tau^{0.496}}\right] \tag{4}$$

According to Reference [35], the creep degree of concrete is taken as follows:

$$C(t,\tau) = \left(f_1 + g_1\tau^{-P_1}\right)\left[1 - e^{-\tau_1(t-\tau)}\right] + \left(f_2 + g_2\tau^{-P_2}\right)\left[1 - e^{-\tau_2(t-\tau)}\right] \tag{5}$$

$$\text{Adiabatic temperature rise } (^\circ\text{C}): \ \theta(\tau) = 56.1\tau/(\tau + 1.76) \tag{6}$$

3.3. Other Thermal Parameters

The numerical value of the surface heat release coefficient of concrete in the air is related to the wind speed. According to engineering data, the average wind speed in the construction area over the years was 2.0 m/s. Since the surface of concrete is smooth after demolding, the value is taken based on the smooth surface as follows.

When there is wind [35],

$$\beta = 18.46 + 17.36v_a^{0.883} = 18.46 + 17.36 \times 2.09^{0.883} = 51.74 \text{ kJ}/\left(\text{m}^2\cdot\text{h}\cdot^\circ\text{C}\right). \tag{7}$$

The surface of the foundation is considered rough and can be divided thus:

$$\beta = 21.06 + 17.58v_a^{0.910} = 21.06 + 17.58 \times 2.09^{0.910} = 55.44 \text{ kJ}/\left(\text{m}^2\cdot\text{h}\cdot^\circ\text{C}\right) \tag{8}$$

4. Calculation Model and Boundary Conditions

4.1. Calculation Model

To ensure the accuracy of the calculation results, it is necessary to establish a calculation model that is as consistent as possible with the actual engineering situation. Considering the pouring process during the construction period, the entire pump station is divided into left and right sections, which are poured separately. The left and right sections have similar structures and are separated by transverse joints. They are poured at the same time. Therefore, in the simulation calculation analysis, the left section is taken as the research object for analysis.

Considering the efficiency of multi-field coupling simulation, this study only uses half of the symmetrical finite element model for simulating the pump station structure. The computational model mesh is shown in Figure 1, and the simulation range includes the pump station and the foundation within a certain range. The model is divided into a total of 137,209 three-dimensional hexahedral elements and 179,444 nodes. The Cartesian coordinate system used for calculation is defined thus: in addition to being parallel to the water, the flow direction is the X-axis, the dam axis is the Y-axis, and the elevation direction is the Z-axis.

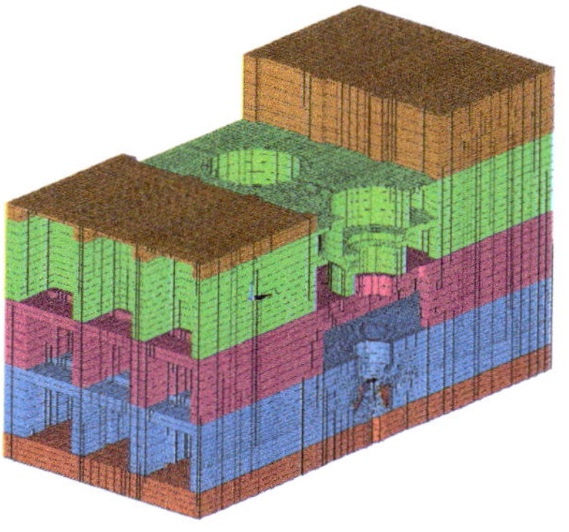

Figure 1. Finite element calculation model of the pump station.

4.2. Feature Points and Cross-Sections

In order to facilitate the observation of the maximum temperature and maximum tensile stress inside the concrete structure of the pump station, the maximum temperature and maximum tensile stress at each time within the time range have been calculated and concentrated into one graph each, namely the temperature envelope graph and the stress envelope graph. The characteristic section with high stress has been sliced (middle section), and the specific location of the characteristic section of the pump station is shown in Figure 2. The feature point selection model is located near the actual temperature measurement point, with the coordinate origin located at the center of the first hole on the baseplate of the left inlet channel of the pump station. The specific coordinates of the feature points are shown in Table 2.

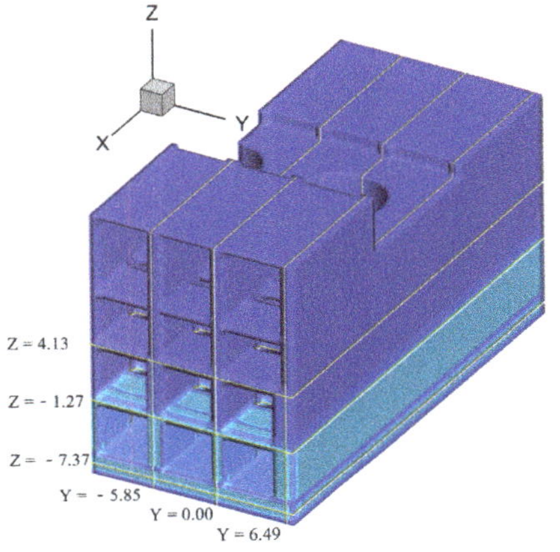

Figure 2. Typical section layout.

Table 2. Coordinates of the feature points (m).

Location Description	Coordinates		
	X	Y	Z
Baseplate surface points	−19.55	−19.55	0.00
Internal points of the baseplate	−6.37	−7.37	−4.84

It should be noted that considering the efficiency of multi-field coupling simulation calculations, this paper only conducts thermo-mechanical-fatigue-damage multi-field coupling simulation calculations on the symmetrical finite element model of the pump station structure. The selection of characteristic sections is half of the overall finite element model, and the selection of feature points is consistent with the overall finite element model.

4.3. Initial and Boundary Conditions

Temperature boundary: The surface of the foundation serves as the heat dissipation surface while the bottom and sides are considered adiabatic boundaries. The symmetry plane of the thermo-mechanical-fatigue-damage multi-field coupling simulation calculation is the adiabatic boundary. Starting from the pouring date, we calculate the temperature field of the foundation 30 years ago to obtain the temperature of the foundation on the pouring date. The heat dissipation of each surface of concrete is determined according to different schemes.

Stress boundary: The bottom of the foundation is fully constrained and the surrounding areas are subjected to normal constraints. The symmetry plane of the thermo-mechanical-fatigue-damage multi-field coupling simulation is subjected to normal constraints.

4.4. Pouring Temperatures

According to the pouring plan, the pump station baseplate will be poured from the end of May to June. Considering the ambient temperature at that time and the measured warehousing temperature, 25 °C will be selected as the initial pouring temperature for the baseplate pouring for calculation and analysis. Table 3 shows the pouring temperatures for each layer of the pump station.

Table 3. Actual pouring temperature of each layer of the pump station.

Serial Number	Pouring Layer	Pouring Temperature (°C)	Start Pouring Time
1	Cushion	19	May 2023
2	Baseplate	25	June 2023
3	Inlet channel layer (Elevation −6.37~−0.77 m)	26	September 2023
4	Outflow channel layer (Elevation −0.77~4.63 m)	21	October 2023
5	Mechanical and electrical layer (Elevation 4.63~9.3 m)	14	November 2023
6	Above an elevation of 9.3 m	10	December 2023

5. Model Establishment, Dynamic Optimization, and Validation

This section is mainly based on using the power exponent function damage constitutive model to establish a concrete thermo-mechanical-fatigue-damage model. The finite element mesh model of the pump station structure is subjected to the multi-field coupled simulation calculation of thermo-mechanical-fatigue-damage. Based on actual monitoring data, various parameters such as the thermal parameters of the material (mainly adiabatic temperature rise in this article) are dynamically optimized and inverted to obtain a concrete structure prediction model. In addition, based on the calculation results of the concrete thermo-mechanical fatigue damage prediction model from the construction period to the operation period of the pump station concrete structure, a comparative analysis is conducted of the measured and calculated values of the monitoring data of the bottom plate of the key parts of the pump station to verify the rationality of the calculation results of the established thermo-mechanical-fatigue-damage model.

5.1. Model Establishment

5.1.1. Temperature Field Parameters

The occurrence of fatigue damage in concrete can have an impact on the thermal conductivity coefficient of concrete. Therefore, establishing a relationship between the thermal conductivity coefficient of concrete and fatigue damage can help make the calculation of temperature field more in line with the actual situation. According to Fourier's law of the thermal conductivity coefficient [37], the thermal conductivity coefficient of fatigue damage variable D is considered as follows [25,26]:

$$\lambda(D) = \lambda_0(1 - D) \tag{9}$$

Here, $\lambda(D)$ is the effective thermal conductivity coefficient of concrete, when fatigue damage D occurs, in W/(m·K); D is the fatigue damage variable, such that $0 \leq D \leq 1$; and λ_0 is the initial thermal conductivity coefficient of concrete in W/(m·K).

Because the calculation of the thermal conductivity coefficient already takes into account the influence of fatigue damage, the above formula can be used to calculate the thermal diffusivity without considering the influence of fatigue damage.

The temperature field mainly considers the influence of damage on the thermal conductivity coefficient, and the specific values of other parameters in the temperature field can be found in Sections 3.2 and 3.3.

5.1.2. Stress Field Parameters

The elastic modulus of concrete will continuously deteriorate with the accumulation of fatigue damage. According to the "strain equivalence assumption" of fatigue damage, the variation process of the elastic modulus of concrete with age under fatigue damage can be expressed as follows [26]:

$$E(\tau,\ D) = E_0\left[1 - e^{-a\tau^b}(1 - D)\right] \tag{10}$$

Here, D is the fatigue damage variable, $\overline{E}$ is the elastic modulus considering fatigue damage in GPa, and E is the elastic modulus of the non-destructive material in GPa.

The stress field mainly considers the influence of damage on the elastic modulus, and the specific values of other parameters of the stress field have been referred to in Sections 3.2 and 3.3.

In summary, the multi-field coupled calculation process of thermo-mechanical fatigue damage in concrete is shown in Figure 3.

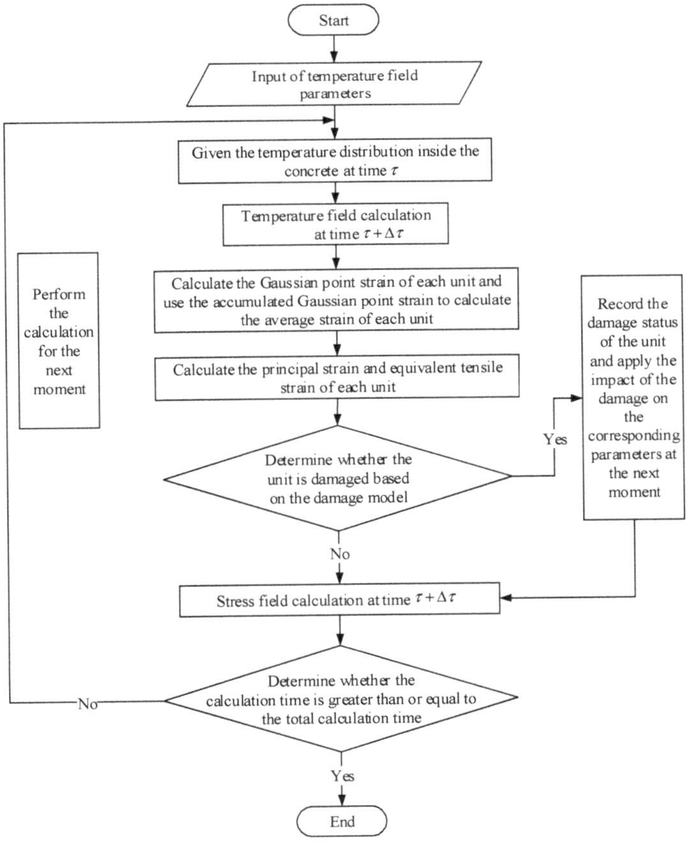

Figure 3. Flowchart for multi-field coupling calculation of thermo-mechanical fatigue damage in concrete.

5.2. Dynamic Optimization

Considering the limited actual monitoring data and the complexity and computational efficiency of multi-field coupled multi parameter inversion, this section only inverts the adiabatic temperature rise of the pump station structure after one year of pouring. The inversion method adopts the classical genetic algorithm [38]. Due to its relatively mature basic theory and method, it will not be further elaborated here.

Based on the dynamic inversion results of actual monitoring data, the optimized fitting formula for the adiabatic temperature rise and age of pump station concrete is obtained as shown in Equation (11).

$$\theta(\tau) = 56.3\tau/(\tau + 1.69) \quad (11)$$

5.3. Model Validation

Based on the dynamic inversion results of actual monitoring data, the adiabatic temperature rise parameters were optimized, and a multi-field coupling prediction model for thermo-mechanical fatigue damage in concrete structures (hereinafter referred to as "Model 2") was ultimately established. In order to verify the rationality of the calculation results of the established thermo-mechanical-fatigue-damage model, Figures 4–7 show the temperature simulation calculation values of the feature points on the baseplate of the inlet channel of Model 2 and the temperature values measured by the embedded detection equipment over time. As mentioned earlier, on the one hand, considering the monitoring time limit, and on the other hand, due to the more significant changes in concrete damage during the construction period compared to the operation period, the actual measurement data of concrete during the construction period (covering temperature, displacement, damage, and stress) are presented separately for 30 days in this article while the corresponding data during the operation period are not listed separately. In contrast, the actual measurement data points during the construction period are more dense, which is more conducive to the observation and comparative analysis involving the actual measurement data and calculated values during the operation period. The coverage and time span of the actual measurement data comprise about one year. Figures 4–7 also add the temperature simulation calculation values of the feature points on the baseplate of the inlet channel without considering fatigue damage (hereinafter referred to as "Model 1"), and comprehensively compare and analyze the above data, as follows.

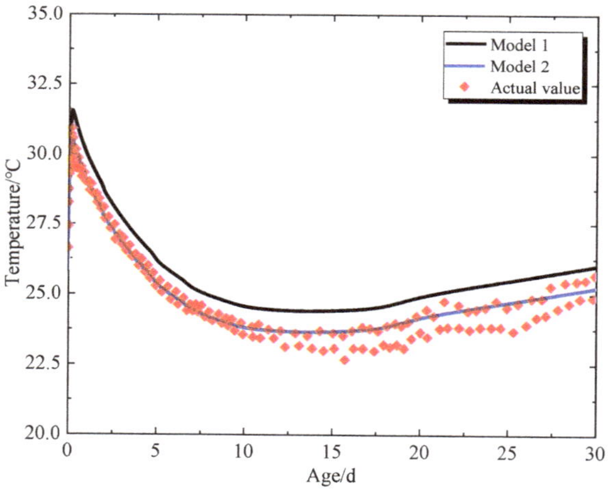

Figure 4. Temperature history of feature points on the surface of the baseplate during the early stage of construction (within 30 days of pouring).

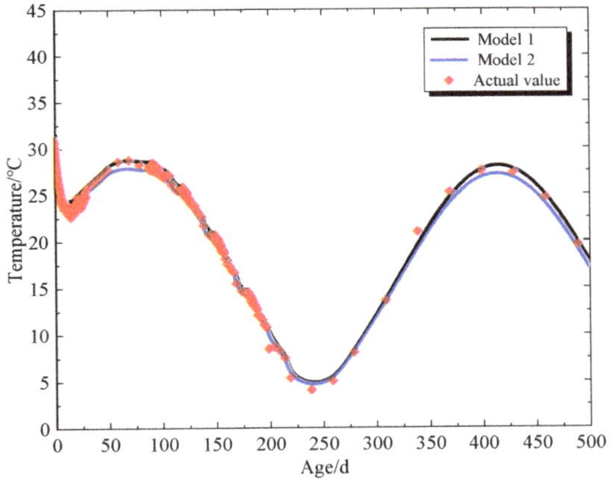

Figure 5. Temperature history of feature points on the surface of the baseplate during the construction period (1 year after pouring).

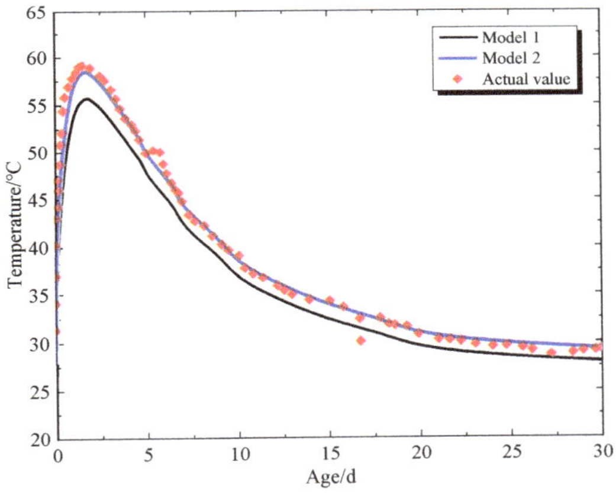

Figure 6. Temperature history of internal feature points of the baseplate during the early stage of construction (within 30 days of pouring).

As shown in Figures 4–7, in terms of surface temperatures, considering the effect of fatigue damage, the calculated values of surface temperatures in the early stage of baseplate construction decreased compared to Model 1 (without considering fatigue damage), with the maximum differences being about 1 °C and 2 °C from the measured values, respectively. Compared to the calculated results of Model 1 (without considering fatigue damage), the maximum errors had decreased by about 1 °C and 1 °C, respectively. During operation, the calculated value of the baseplate surface temperature was also closer to the measured value compared to Model 1. In terms of internal temperature, considering the effect of fatigue damage, the calculated values of internal temperature in the early stage of baseplate construction had increased compared to Model 1 (without considering fatigue damage), with the maximum differences being about 2 °C and 1.5 °C from the measured values, respectively. Compared to the calculated results of Model 1 (without considering fatigue damage), the maximum errors had decreased by about 1 °C and 0.5 °C, respectively. During operation, the calculated temperature inside the baseplate was also closer to the measured value compared to Model 1.

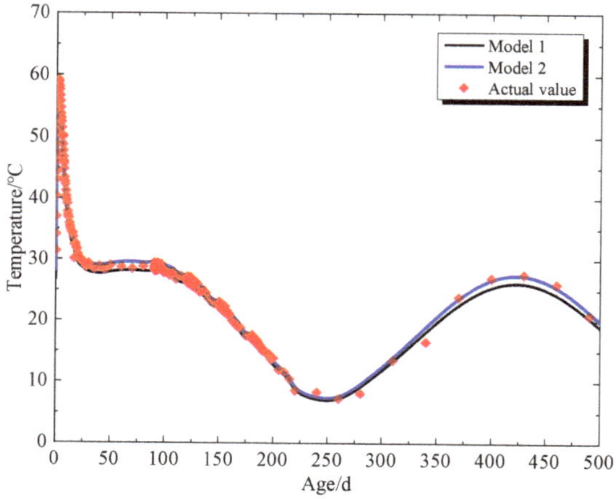

Figure 7. Temperature history of internal feature points of the baseplate during the construction period (1 year after pouring).

Overall, considering the effect of fatigue damage, the degree of agreement between the measured and calculated values of the surface and internal temperatures of the pump station baseplate was relatively higher compared to Model 1 (without considering fatigue damage), with the maximum errors reduced by about 1 °C and 1 °C, respectively. This indicates that there are varying degrees of fatigue damage to the concrete structure of a pump station during both the construction and operation periods. Based on the previous analysis, the following conclusion has been further verified: fatigue damage can lead to a decrease in the internal thermal conductivity coefficient of concrete, manifested macroscopically as a slower rate of heat conduction from the inside to the surface; insufficient time for internal heat to diffuse to the surface, resulting in heat accumulation and increase; and the insufficient timely replenishment of internal heat to reduce the surface temperature.

In summary, fatigue damage can affect the thermal conduction process of concrete, thereby affecting the temperature distribution of concrete. At the same time, the calculation results of the established thermo-mechanical-fatigue-damage model have been verified to be reasonable and feasible. Therefore, in the simulation calculation process of concrete structure construction and operation, it is necessary to consider the impact of fatigue damage on the temperature, deformation, stress, etc. of the concrete structure, and then the simulation calculation results can more accurately predict the actual temperature, deformation, stress, and other changes.

6. Analysis of the Results

6.1. Analysis of the Influence of Fatigue Damage

The main purpose of this section is to clarify the performance response parameters of the pump station during the construction operation period, including the stress performance, deformation performance, and dynamic fatigue damage, based on the established concrete thermo-mechanical fatigue damage prediction model under the existing construction conditions and construction progress. The influence analysis of the temperature field, deformation, and stress field of the calculation results is also conducted.

Figures 8–16 correspond to the envelope diagrams of the highest temperature, maximum displacement, maximum damage, and maximum stress in various parts of the pump station one year after the start of pouring the baseplate in May 2023 at a temperature of 25 °C (as shown in Section 4.4).

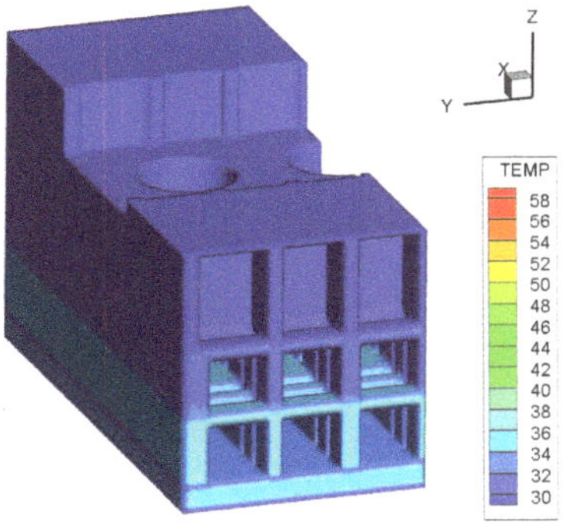

Figure 8. Envelope diagram of the highest temperature on the left side surface (1 year after pouring) in °C.

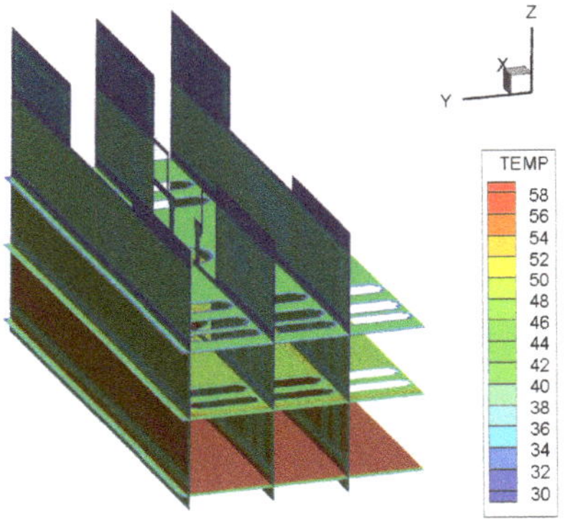

Figure 9. Envelope diagram of the highest temperature in the cross-section of each layer in the left side (1 year after pouring) in °C.

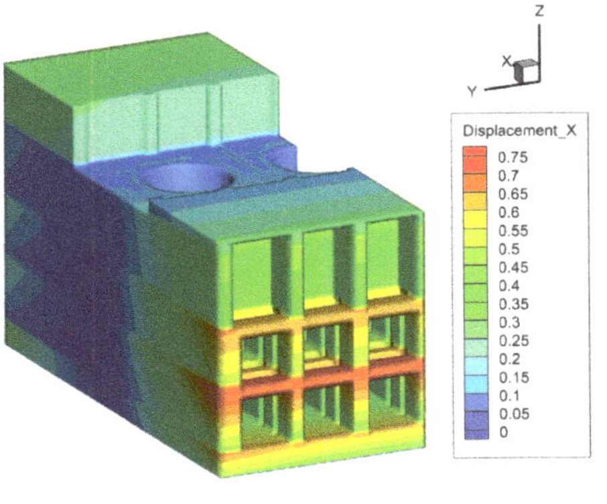

Figure 10. Maximum displacement in the direction of water flow on the left side (1 year after pouring) in cm.

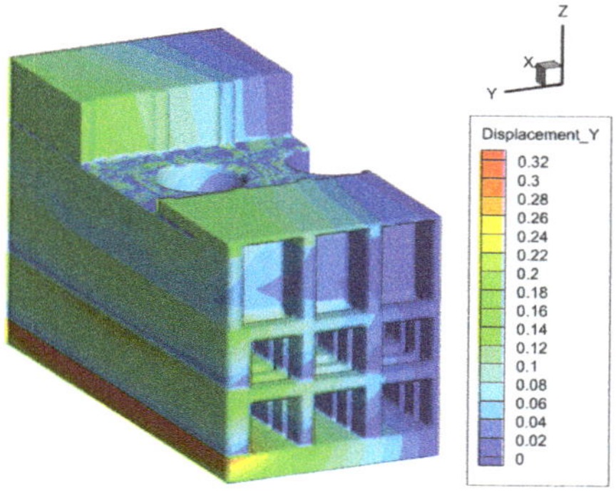

Figure 11. Maximum displacement contour in the vertical direction of water flow on the left side (1 year after pouring) in cm.

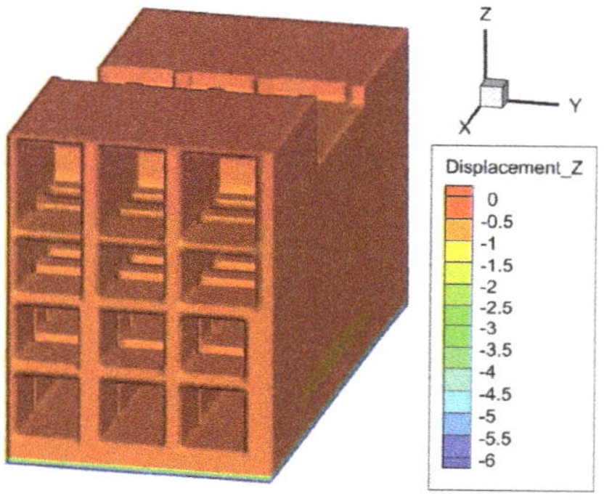

Figure 12. Maximum vertical displacement contour on the left side (1 year after pouring) in cm.

In terms of surface temperatures, as shown in Figure 8, when pouring from the bottom to the top, the peak surface temperatures of individual layers in the upper part show the same variation patterns with a change in the seasonal environmental temperature. Under the effect of fatigue damage, the overall surface temperatures of the pump station decrease relatively, and the highest surface temperatures of individual parts of the pump station decrease by 0.1 °C to 1.0 °C. The maximum temperature on the surface of the pump station baseplate is around 31 °C and the maximum temperature on the surface of the inlet channel pier wall is around 30 °C. The peak temperature of the baseplate surface of the inlet channel layer is around 33 °C. The peak temperature of the baseplate surface of the outflow channel layer is around 33 °C and the peak temperature of the pier wall surface of the outflow channel layer is around 29 °C. The peak temperature of the baseplate surface of the electromechanical layer is around 33 °C and the peak temperature of the top surface of the pump station is around 28 °C.

Overall, during the entire process from construction to operation and under the influence of fatigue damage, the surface temperature of the pump station slightly decreases by 0.1 °C to 1.0 °C. This was mainly due to the fact that the damage causes a decrease in the internal thermal conductivity coefficient of the concrete, manifested macroscopically as a slower thermal conductivity coefficient from the inside to the surface, and the surface temperature does not receive the timely replenishment of internal heat, resulting in a decrease.

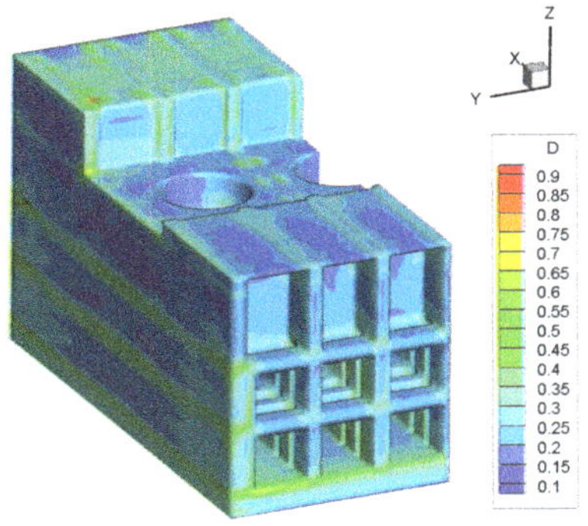

(**a**)

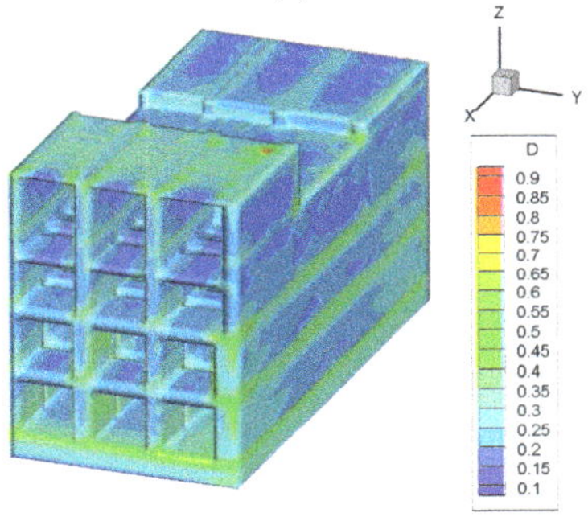

(**b**)

Figure 13. Maximum damage contours on the left side surface (1 year after pouring). (**a**) Upstream side; (**b**) downstream side.

In terms of internal temperature, as shown in Figure 9, considering the effect of fatigue damage, the highest temperatures inside individual parts of the pump station increase by 0.5 °C to 3 °C. Due to the large thickness of the baseplate and single-sided heat dissipation, the peak temperature inside the baseplate reaches around 58 °C, and the peak temperature inside the inlet channel pier wall reaches around 55 °C. The peak temperature inside the baseplate of the inlet channel reaches around 48 °C. The pouring temperature of the baseplate of the outlet channel is 4 °C lower than that of the pump station baseplate, and the thickness is 1 m, which is only half of the thickness of the pump station baseplate and has double-sided heat dissipation. The peak temperature is only about 48 °C while the peak temperature inside the outlet channel pier wall reaches about 49 °C. The upper part of the electromechanical layer pier wall, which is structurally similar, has a peak temperature of about 50 °C.

Overall, during the entire process from construction to operation and under the influence of fatigue damage, the internal temperatures of the pump station slightly increase by 0.5 °C to 3 °C. This is mainly due to the fact that the damage causes a decrease in the thermal conductivity coefficient inside the concrete, which manifests macroscopically as a slower thermal conductivity coefficient rate from the inside to the surface. The internal heat does not have time to diffuse to the surface, resulting in heat accumulation and an increase.

In terms of maximum deformation, as shown in Figures 10–12, most parts of the pump station structure have relatively large displacement differences in the three directions of water flow, vertical water flow, and vertical direction. After considering the fatigue damage effect, the overall displacement of the pump station in the direction of water flow is positive while the displacement in the direction of vertical water flow and vertical direction is negative, that is, downward displacement. The maximum displacements of the pump station structure in the three directions of water flow, vertical water flow, and vertical direction are 0.75 cm, 0.31 cm, and −1.83 cm, respectively, located on the upstream side of the inlet channel baseplate, the left side of the baseplate in the direction of water flow, and the downstream side of the baseplate.

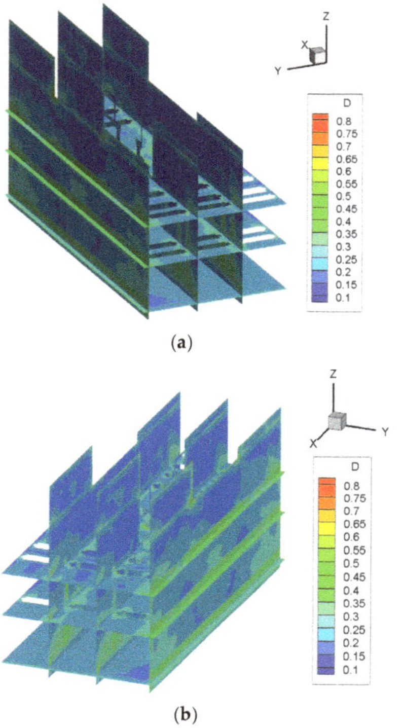

Figure 14. Maximum damage envelope contours inside the left joint (1 year after pouring). (**a**) Upstream side; (**b**) downstream side.

Overall, the vertical displacement values on the surface of the pump station are greater than those along and perpendicular to the water flow direction. The main reason is that the foundation is relatively soft, and the overall settlement during the construction period of the pump station structure is relatively large compared to the displacement caused by temperature loads. Due to the consideration of fatigue damage, the overall temperature difference between the inside and outside of the pump station increases relatively, and the maximum displacements in all directions of the pump station change by −0.21 cm to 0.23 cm relative to Model 1.

In terms of surface damage, as shown in Figure 13, during the entire process from construction to operation of the pump station structure and under fatigue damage, the maximum damage to the pump station surface mainly occurs on the baseplate surface (damage of about 0.47) and the outer surface of the side wall (damage of 0.51–0.81). It should be noted that considering an initial damage of 0.15, the larger the damage value (maximum value of 1) is, the more microcracks are generated in the concrete, the lower the structural strength is, and the lower the safety is.

As shown in Figure 14, in terms of internal damage, during the entire process from construction to operation of the pump station structure and under fatigue damage, the damage to the baseplates on both sides of the pump station is above 0.26, the damage to other areas of the baseplate is around 0.23, the damage to the inlet channel baseplate is around 0.51, the damage to the outlet channel baseplate is around 0.42, the maximum damage to the inlet channel pier wall is between 0.29 and 0.44, the maximum damage to the outlet channel pier wall is between 0.34 and 0.51, and the maximum damage to the mechanical and electrical layer pier wall is between 0.31 and 0.46.

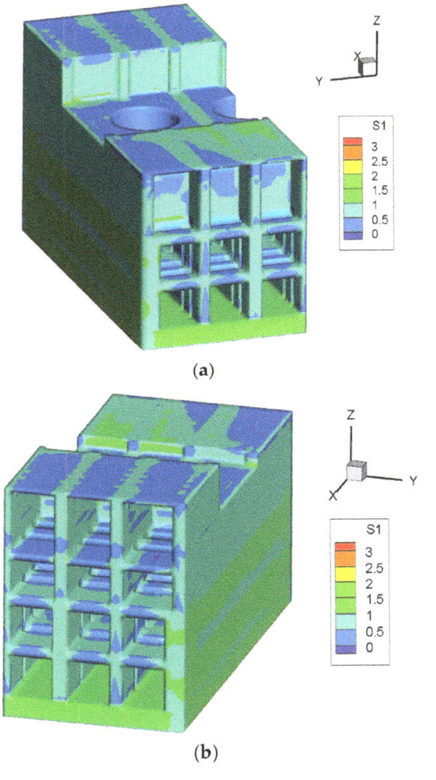

Figure 15. Maximum principal stress contours on the left side surface (1 year after pouring) in MPa. (**a**) Upstream side; (**b**) downstream side.

As shown in Figure 15, in terms of surface stress, the maximum surface stress of the pump station structure during the entire process from construction to operation and under fatigue damage mainly occurs on the baseplate surface at about 1.6 MPa and the outer surface of the side walls at 1.1~1.4 MPa.

As shown in Figure 16, in terms of internal stress, during the entire process from construction to operation of the pump station structure and under fatigue damage, the stress on the baseplates on both sides of the pump station is above 1.5 MPa, the stress on other areas of the baseplate is around 1.3 MPa, the stress on the baseplate of the inlet channel is around 1.5 MPa, the stress on the baseplate of the outlet channel is around 1.6 MPa, the maximum stress on the inlet channel pier wall is between 2.2 MPa and 2.7 MPa, the maximum stress on the outlet channel pier wall is between 1.6 MPa and 1.9 MPa, and the maximum stress on the mechanical and electrical layer pier wall is between 1.8 MPa and 2.2 MPa.

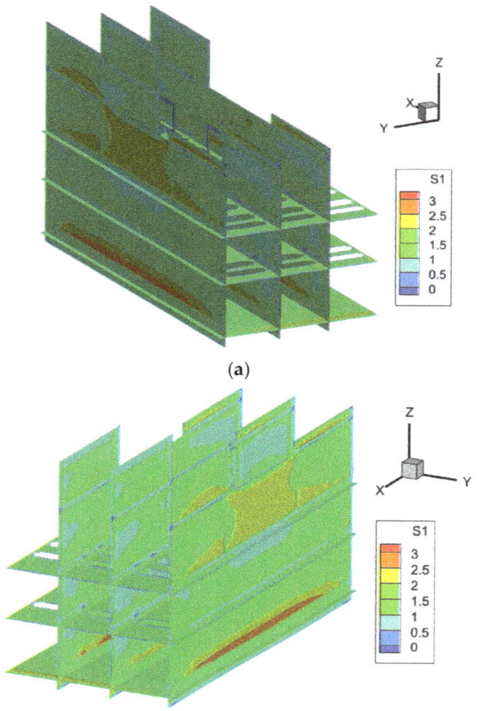

(**a**)

(**b**)

Figure 16. Maximum principal stress envelope contours inside the left side (1 year after pouring) in MPa. (**a**) Upstream side; (**b**) downstream side.

6.2. The Variation Law of Characteristic Parameters over Time

The main purpose of this section is to analyze the calculation results based on the established concrete thermo-mechanical fatigue damage prediction model and clarify the temperature, deformation, fatigue damage, and stress development and variation laws of typical parts of the pump station from the construction period to the operation period.

6.2.1. Time Dependent Variations of Characteristic Parameters

Figures 17–26 show the temperature, displacement, stress history, and temperature difference changes of the feature points on the baseplate of the inlet channel. The surface feature points are located on the surface of the baseplate and the internal feature points are located at the center elevation of the baseplate. The specific coordinate positions are shown in Section 4.2. It should be noted that the temperature simulation calculation values and actual monitoring values have been analyzed in Section 5.2, and no comparative analysis will be conducted in this section. The focus will be on analyzing the temperature, displacement, fatigue damage, and stress changes over time in typical parts of the pump station.

As shown in Figures 17 and 18, in terms of surface temperature, the surface points reach a peak temperature of 30.67 °C within 2 days after pouring, displaying a decrease of about 1.0 °C compared to Model 1. This is mainly due to the consideration of fatigue damage. The internal thermal conductivity coefficient of the bottom concrete will decrease, which is manifested macroscopically as a slower heat conduction rate from the inside of the bottom to the surface, and the surface temperature will not receive the timely replenishment of internal heat, resulting in a decrease. Afterwards, due to the heat dissipation effect of the bottom surface, the overall temperature of the bottom concrete decreases to near the ambient temperature with the decrease in temperature. During operation, due to the slow development of the elastic modulus of the bottom concrete and its high strength, the fatigue damage caused by changes in ambient temperature to the bottom concrete is relatively small. However, the temperature of the bottom concrete changes periodically with changes in ambient temperature during operation.

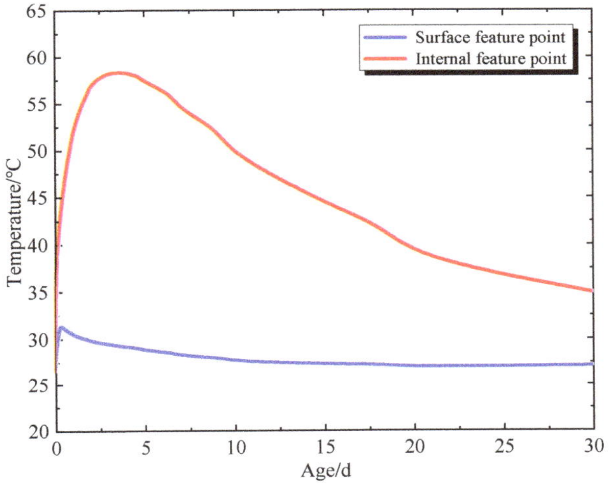

Figure 17. Temperature histories of the feature points on the baseplate during the early stage of construction (within 30 days of pouring).

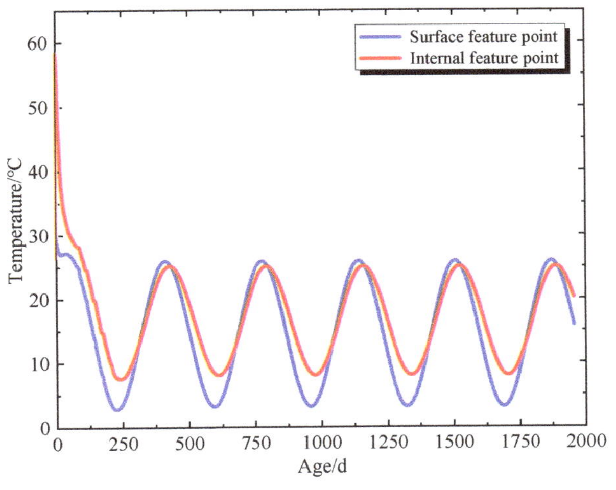

Figure 18. Temperature histories of feature points on the baseplate during the operation period (1 year after pouring).

As shown in Figures 17 and 18, in terms of internal temperature, considering the effect of fatigue damage, the peak temperature inside the bottom plate reaches 58.6 °C about 3 days after pouring, which has increased by about 3 °C compared to Model 1. This is mainly due to the decrease in thermal conductivity coefficient inside the bottom plate concrete after considering fatigue damage, which is manifested macroscopically as a slower heat conduction rate from the inside of the bottom plate to the surface, and the internal heat does not have time to diffuse to the surface, resulting in heat accumulation and increase. Afterwards, due to the heat dissipation effect of the top surface of the bottom plate, the temperature of the bottom plate gradually decreases. By February 2024, the internal temperature of the bottom plate has decreased to about 6 °C, and the temperature of the bottom plate concrete has also shown periodic changes with the ambient temperature during operation.

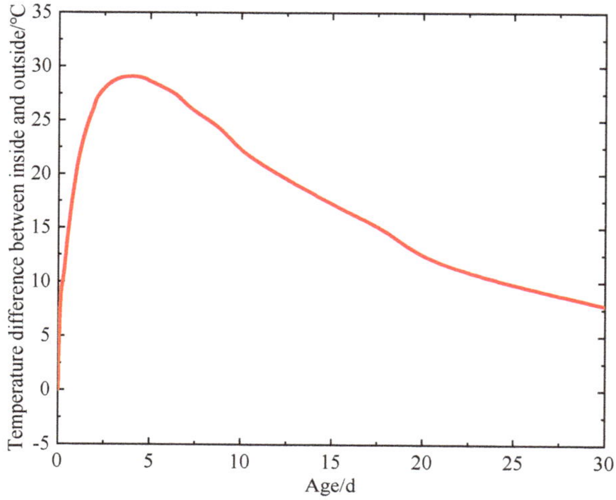

Figure 19. Early construction process of temperature difference inside and outside the baseplate (within 30 days of pouring).

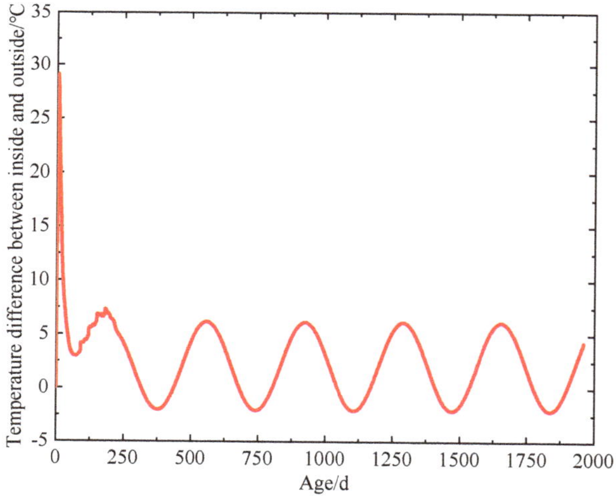

Figure 20. Construction period history of temperature difference inside and outside the baseplate (1 year after pouring).

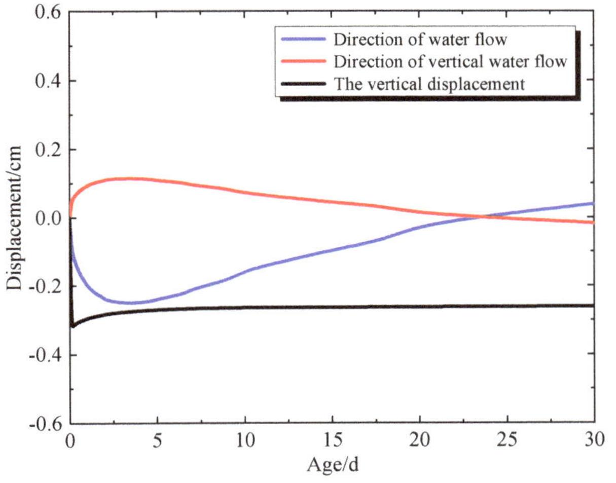

Figure 21. Maximum displacements during early stage of baseplate construction (within 30 days of pouring).

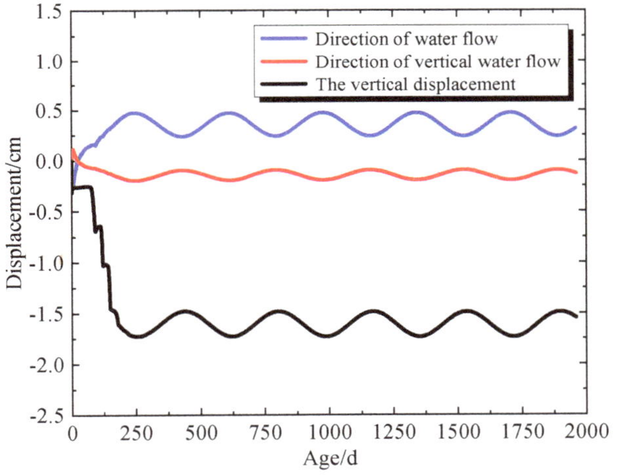

Figure 22. Maximum displacements during the construction period of the baseplate (1 year after pouring).

As shown in Figures 19 and 20, in terms of temperature difference between the inside and outside, about 3 days after pouring, the maximum temperature difference between the inside and outside of the bottom plate reaches 30.5 °C, which has increased by about 3 °C compared to Model 1. Afterwards, as the bottom plate surface dissipates heat, the temperature difference between the inside and outside gradually decreases. By February 2024, the temperature difference between the inside and outside of the bottom plate has decreased to within 5 °C, and the temperature difference between the inside and outside of the concrete of the bottom plate has also shown periodic changes with the ambient temperature during operation.

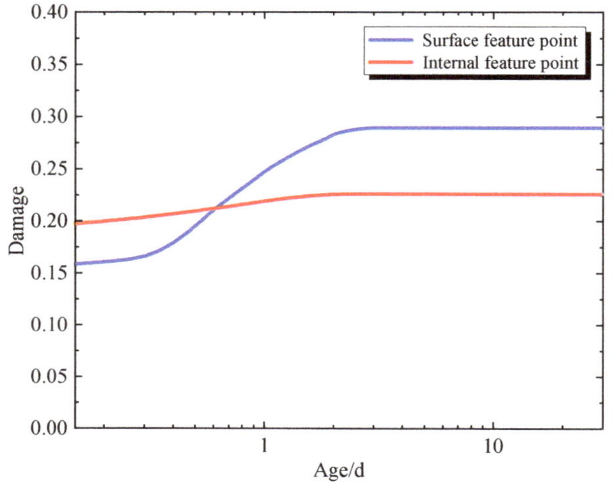

Figure 23. Damage processes of the feature points on the baseplate during early construction (within 30 days of pouring).

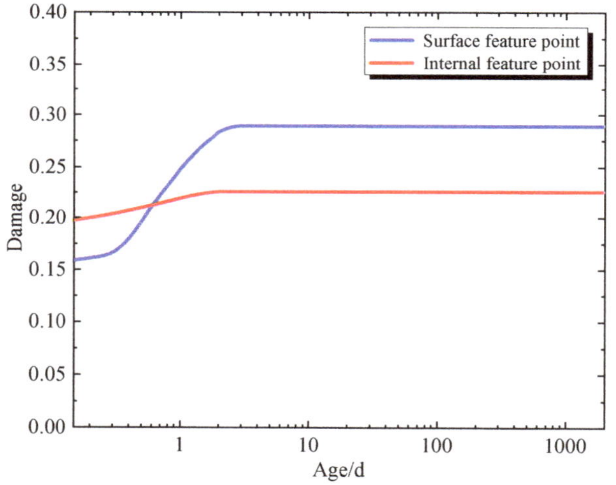

Figure 24. Damage histories of the feature points on the baseplate during operation (1 year after pouring).

In terms of maximum deformation, as shown in Figures 21 and 22, there is not much difference between the displacement of the baseplate and the surface displacement with an increasing time. After pouring, the displacement of the baseplate in the direction of water flow first decreases with an increasing time and then increases to around 0.48 cm. The displacement in the direction of vertical water flow does not change significantly with the increase in time, with a maximum displacement of about 0.11 cm. The vertical displacement initially does not change significantly with the increase in time, but later, under the influence of the gravity of the upper pouring layer, the displacement gradually decreases to around −1.74 cm and remains basically unchanged. During the operation period, the maximum displacement inside the concrete of the baseplate is periodically affected by the ambient temperature.

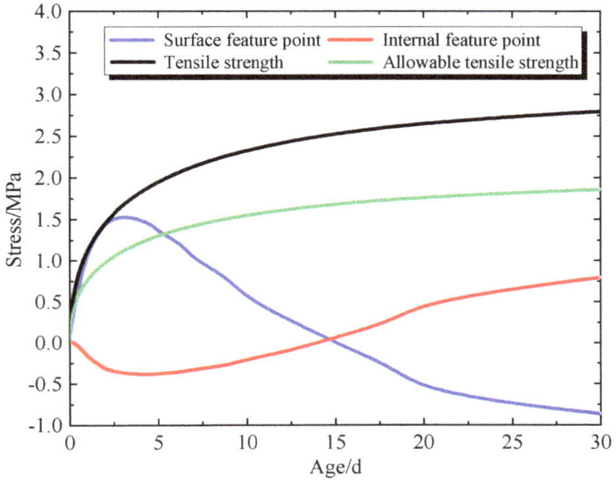

Figure 25. Stress histories of the feature points on the baseplate during the early stage of construction (within 30 days of pouring).

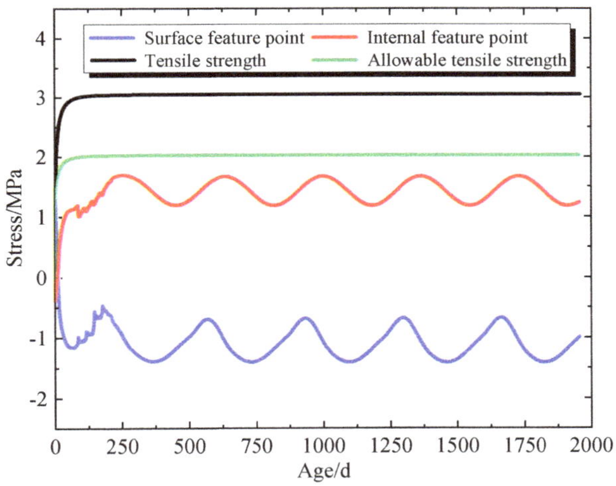

Figure 26. Stress histories of the feature points on the baseplate during operation (1 year after pouring).

After considering the effect of fatigue damage, the peak surface temperature of the baseplate decreases by about 1.0 °C, the peak internal temperature increases by about 3 °C, and the temperature difference between the inside and outside increases by about 3 °C. The overall maximum displacements of the baseplate in all directions relative to Model 1 vary by −0.16 cm to 0.07 cm. Due to the slow development and high strength of the elastic modulus of the bottom concrete during operation, the fatigue damage caused by changes in the ambient temperature to the bottom concrete is relatively small. However, the maximum displacement of the bottom concrete varies periodically with changes in the ambient temperature during operation.

As shown in Figures 23 and 24, in terms of surface damage, due to the large temperature difference between the inside and outside (27 °C), the surface damage of the baseplate rapidly increases to around 0.28 within 3 days of pouring. During the operation period, due to the slow development of the elastic modulus of the baseplate concrete and its high strength, the fatigue damage caused by changes in the ambient temperature to the baseplate concrete is relatively small, and the growth of damage during the operation period is slow until it remains unchanged. In terms of internal damage, the internal damage of the baseplate also rapidly increases to around 0.23 within 3 days of pouring. During the same operation period, the internal damage of the baseplate grows slowly and remains unchanged, but the overall surface damage of the baseplate is slightly reduced. After considering the effect of fatigue damage, the early surface damage of the baseplate increases by about 0.15 relative to Model 1 (initial damage of 0.15), and the internal damage of the baseplate increases by about 0.10 (initial damage of 0.15). During operation, the growth of baseplate damage slows down and remains unchanged.

As shown in Figures 25 and 26, in terms of surface stress, due to the large temperature difference between the inside and outside (28 °C), the surface of the baseplate exceeds the allowable tensile strength of the corresponding age concrete after 2 days of pouring (about 2 days of age), with a tensile stress of 1.53 MPa, and the risk of concrete cracking is extremely high. As the overall temperature of the baseplate decreases in the later stage, the temperature difference between the inside and outside decreases and the tensile stress on the surface of the baseplate gradually decreases. During operation, due to the slow development of the elastic modulus of the baseplate concrete and its high strength, the fatigue damage caused by changes in the ambient temperature to the concrete is relatively small, and the impact of damage on stress is also relatively small. During operation, the baseplate surface will experience stress fluctuations with a range of −1.4 MPa to −0.5 MPa due to seasonal changes.

In terms of internal stress, the early temperature hydration heat causes the internal compression of the baseplate, and as the temperature increases, the compressive stress gradually increases. After the internal temperature of the baseplate reaches its peak and enters the cooling period, the internal stress begins to gradually transform from compressive stress to tensile stress. During the wintering process, the maximum tensile stress reaches around 1.7 MPa (around February 2024), which is lower than the tensile strength. Similarly, due to seasonal changes during operation, stress fluctuations ranging from 1.2 MPa to 1.7 MPa will occur inside the baseplate.

Overall, considering the effect of fatigue damage, the temperature difference between the inside and outside of the baseplate increases, leading to stress growth. However, damage can also cause a decrease in the elastic modulus of the baseplate, resulting in a decrease in stress. Under the combined effect of the two, the specific manifestation is that the early surface tensile stress of the baseplate increases by about 0.1 MPa relative to Model 1 and the internal tensile stress of the baseplate decreases by about 1.0 MPa. During the overwintering period, the surface tensile stress increases by about 0.1 MPa while the internal tensile stress decreases by about 1.5 MPa. During the operation period, due to seasonal changes, stress fluctuations with amplitude ranges of −1.4 MPa to −0.5 MPa and 1.2 MPa to 1.7 MPa will occur on the surface and inside of the baseplate, respectively. The above fatigue damage will cause more microcracks in the baseplate structure, which is extremely detrimental to the safety of the baseplate structure during operation.

6.2.2. Peak Values of Various Parts

Table 4 summarizes the maximum stresses, maximum displacements, and peak temperatures of feature points at different locations of the pump station. Based on the simulation results of the fatigue damage model considering the actual pouring layers of the pump station, the tensile stresses of the feature points of the pump station baseplate poured in May and the pier wall poured in September are relatively reduced. However, the early stress of the pier wall will still exceed the allowable tensile strength of the concrete, and even the tensile strength, and the risk of cracking is still high, mainly manifested as the easy occurrence of surface cracks in the early stage of the pier wall.

Table 4. Peak temperatures, maximum deformations, maximum damage, and stresses at the feature points of pump station.

Feature Location	Feature Point	Temperature Peak/°C	Maximum Displacement/cm			Maximum Damage	Maximum Principal Stress/MPa
			X	Y	Z		
Baseplate	Surface	30.67	0.48	0.11	−1.74	0.29	1.53
	Internal	58.6				0.23	1.71
Inlet channel pier wall	Surface	29.79	0.25	0.06	−1.74	0.44	1.21
	Internal	55.57				0.23	2.63
Inlet channel baseplate	Surface	32.53	0.72	0.15	−1.41	0.44	0.8
	Internal	47.89				0.51	2.09
Outlet channel pier wall	Surface	32.51	0.38	0.16	−1.78	0.31	0.61
	Internal	47.81				0.45	1.41
Outlet channel baseplate	Surface	30.68	0.27	0.13	−1.4	0.81	1.12
	Internal	51.73				0.51	2.21
Mechanical and electrical layer pier wall	Surface	34.86	0.37	0.05	−0.79	0.31	1.09
	Internal	49.97				0.38	1.29

6.3. Key Control Structural Parameters, Warning Indicators, and Thresholds

Based on the multi-field coupling numerical simulation prediction and analysis results of thermo-mechanical fatigue damage in the entire process of pump station structure construction operation, the performance response parameters of the structure during operation, such as stress performance, deformation performance, and dynamic fatigue, have been obtained. Combined with the comparison of the results between the calculated values in Section 5.3 and the response characteristics of the solid structure (monitoring values), the following key control structural parameters for the long-term safe operation of the structure have been proposed (mainly the peak characteristic temperatures of key parts, the temperature difference between the inside and outside, the maximum deformation, the maximum damage, and the maximum tensile stress), together with the warning indicators (mainly the peak characteristic temperatures of key parts, the temperature difference between the inside and outside, the maximum deformation, and the maximum tensile stress) and the thresholds (mainly those corresponding to the peak characteristic temperatures of key parts, the temperature difference between the inside and outside, the maximum deformation, and the maximum tensile stress). As shown in Table 5, the ultimate goal is to ensure the long-term safe and efficient operation of the structure.

Table 5. Key control structure parameters, warning indicators, and thresholds for long-term safe operation of pump station structures.

Key Control Structural Parameters	Temperature Peak	Temperature Difference Between Inside and Outside	Maximum Displacement	Maximum Tensile Stress		Maximum Damage
Warning Indicators	Temperature peak	Temperature difference between inside and outside	Maximum displacement	Maximum tensile stress		/
Thresholds	≤55 °C	≤26 °C	≤1.7 cm	Early stage of construction	≤1.5 MPa	0.4
				Operation period	≤2.9 MPa	

This paper is based on the power exponent function damage constitutive model and has established a concrete thermo-mechanical-fatigue-damage model. It has focused on analyzing and exploring the mechanical behavior and damage evolution process in concrete structures under temperature changes from a three-dimensional perspective and analyzed the influences of damage on thermal diffusivity and the thermal conductivity coefficient. However, there has been a lack of research in simulating concrete crack propagation, and it is necessary to carry out relevant research in this area in the next step.

7. Conclusions

According to the actual construction pouring plan of the pump station concrete, and based on the above analysis results, it can be seen that the established thermo-mechanical-fatigue-damage model has calculated the results that are closer to the measured values and have higher prediction accuracy. The performance parameters of the pump station structure, such as stress performance, deformation performance, and dynamic fatigue damage, have been obtained for the construction operation period. The main conclusions are drawn as follows.

(1) Fatigue damage can lead to a decrease in the internal thermal diffusivity of concrete, manifested macroscopically as a slower rate of heat conduction from the inside to the surface; insufficient time for internal heat to diffuse to the surface, resulting in heat accumulation and increase; and the insufficient timely replenishment of internal heat to reduce the surface temperature. That is, the peak surface temperature of the pump station structure decreases by about 1.0 °C, the peak internal temperature increases by about 3 °C, and the temperature difference between the inside and outside increases by about 3 °C.

(2) After considering the effect of fatigue damage, the temperature of the pump station baseplate during the wintering period will still drop to around 5 °C and the tensile stress of the baseplate will rapidly increase with a decrease in the temperature. Although the temperature difference between the inside and outside is relatively high, the internal tensile stress of the baseplate during operation will decrease by about 1.5 MPa. The main reason is that the damage will cause a decrease in the elastic modulus of the concrete, resulting in a decrease in stress. However, the relative temperature difference between the inside and outside will lead to a greater increase in stress, which will be manifested as a decrease in the overall stress of the baseplate. Fatigue damage will cause more microcracks in the pump station structure, thereby reducing the strength of the pump station structure. This is extremely unfavorable for the safety of the pump station structure during operation.

(3) This paper proposes warning indicators and thresholds for key parts of the pump station, including a temperature peak threshold of 55 °C, an internal and external temperature difference threshold of 26 °C, a maximum deformation threshold of 1.7 cm, a maximum damage threshold of 0.4, and thresholds corresponding to maximum tensile stress during early construction and operation periods of 1.5 MPa and 2.9 MPa, respectively, which can provide reference for similar projects.

Author Contributions: Conceptualization, M.Y.; methodology, M.Y. and X.Z.; programming, M.Y.; formal analysis, M.Y.; investigation, M.Y., X.Z., L.H., Y.Q., Z.X., X.Q. and Y.D.; writing—original draft, M.Y. and X.Z.; writing—review and editing, M.Y. and X.Z.; funding acquisition, M.Y., X.Z., Z.X. and X.Q.; data curation, X.Z., L.H., Y.Q., Z.X., X.Q. and Y.D.; resources, L.H., Y.Q., Z.X., X.Q. and Y.D.; supervision, Z.X. All authors have read and agreed to the published version of the manuscript.

Funding: This work was financially supported by the Fundamental Research Funds for Central Public Welfare Research Institutes (Grant Nos. CKSF20241000/CL and CKSF20241023/CL), Anhui Provincial Group Limited for the Yangtze-to-Huaihe Water Diversion Technology Support Project (Grant No. YJJH-ZT-ZX-20210408381), the National Natural Science Foundation of China (Grant No. U2340229), and the Natural Science Foundation of Wuhan (Grant No. 2023020201020360).

Data Availability Statement: Data are contained within the article.

Acknowledgments: The authors acknowledge the financial support from the National Natural Science Foundation of China, the Fundamental Research Funds for Central Public Welfare Research Institutes, Anhui Provincial Group Limited for the Yangtze-to-Huaihe Water Diversion Technology Support Project, and the Natural Science Foundation of Wuhan.

Conflicts of Interest: Authors Xianhong Zhu and Lei Han were employed by the company Anhui Provincial Group Limited for Yangtze-to-Huaihe Water Diversion. The remaining authors declare that the research was conducted in the absence of any commercial or financial relationships that could be construed as a potential conflict of interest. The authors declare that this study received funding from Anhui Provincial Group Limited for the Yangtze-to-Huaihe Water Diversion. The funder was involved in the study design, collection, analysis, interpretation of data, the writing of this article or the decision to submit it for publication.

References

1. Zhang, G.; Zhou, Q. Study on temperature recovery after arch sealing of super high arch dams and its influence. *J. Hydraul. Eng.* **2015**, *46*, 1009–1018. (In Chinese)
2. Xie, Y.; Qian, C.; Xu, Y.; Wei, M.; Du, W. Effect of fine aggregate type on early-age performance, cracking analysis and engineering applications of C50 concrete. *Constr. Build. Mater.* **2022**, *323*, 126633. [CrossRef]
3. Yu, T.; Wu, Y.; Yang, Q.; Yang, M.; Jin, S.; Jiang, Y. Analysis and control of temperature and crack inside and outside mass concrete cavity structure. *J. Railw. Sci. Eng.* **2020**, *17*, 690–698. (In Chinese)
4. Cai, C.; Tang, H.; Wen, T.; Li, J.; Chen, Z.; Li, F.; Su, Q.; Li, R. Long-term shrinkage performance and anti-cracking technology of concrete under dry-cold environment with large temperature differences. *Constr. Build. Mater.* **2022**, *349*, 128730. [CrossRef]
5. Gao, P.; Chen, Y.; Huang, H.; Wei, J.; Yu, Q. Quantification of damage in concrete under drying condition. *J. Chin. Ceram. Soc.* **2021**, *49*, 312–322. (In Chinese)
6. Meyer, D.M.; Boshoff, W.P.; Combrinck, R. Utilising super absorbent polymers as alternative method to test plastic shrinkage cracks in concrete. *Constr. Build. Mater.* **2020**, *248*, 118666. [CrossRef]
7. Afefy, H.M.; El-Sensar, M.S.; Kassem, N.M.; Taher, S.E.-D.F. Structural performance of perforated arched self-compacted reinforced concrete beams under concentrated static loading. *Structures* **2023**, *58*, 105687. [CrossRef]
8. Andreikiv, E.; Dolinska, I.; Raiter, O.K. Computational model for the evaluation of the service life of fiber-reinforced concrete structures under long-term static loading. *Mater. Sci.* **2020**, *56*, 291–300. [CrossRef]
9. Shi, J.; Zhao, Y. Pore structure of concrete under fatigue load in areas with large temperature differences. *Struct. Concr.* **2022**, *23*, 2132–2149. [CrossRef]
10. Cheng, Z.; Wang, W.; Tang, J.; Feng, H.; Zhou, J.; Wang, K. Peridynamic simulation of engineered cementitious composites and concrete laminated structure under fatigue loading. *Constr. Build. Mater.* **2024**, *428*, 136270. [CrossRef]
11. Yuan, M.; Qiang, S.; Xu, Y.; Li, Y.; Xu, W. Research on cracking mechanism of early-age restrained concrete under high-temperature and low-humidity environment. *Materials* **2021**, *14*, 4084. [CrossRef]
12. Zhang, C.; Zhang, J.; Li, H.; Li, Y.; Xu, W.; Qiang, S.; Yuan, M. A phase field approach to the fracture simulation of quasi-brittle structures with initial state. *Eng. Fract. Mech.* **2024**, *304*, 110150. [CrossRef]
13. Jiang, Z.; Liu, F.; Cai, G.; Liu, Z.; Tian, G.; Lu, Z.; Yu, S. A recycled crushed rock sand mortar based on Talbot grading theory: Correlation of pore structure and mechanical properties. *Constr. Build. Mater.* **2024**, *446*, 137980. [CrossRef]
14. Jiang, Z.; Cai, G.; He, H.; Tian, G.; Liu, Y.; Wu, M. Pore structure evolution of mortars with manufactured sand aggregate under the influence of aggregate size and water saturation environment. *Measurement* **2025**, *241*, 115736. [CrossRef]
15. Yuan, M.; Zhou, D.; Chen, J.; Hua, X.; Qiang, S. Study on the calculation method of stress in strong constraint zones of the concrete structure on the pile foundation based on Eshelby equivalent inclusion theory. *Materials* **2020**, *13*, 3815. [CrossRef]
16. Zhu, Z.; Liu, Y.; Fan, Z.; Qiang, S.; Xie, Z.; Chen, W.; Wu, C. Improved buried pipe element method for temperature-field calculation of mass concrete with cooling pipes. *Eng. Comput.* **2020**, *37*, 2619–2640. [CrossRef]
17. Li, Y.; Qiang, S.; Xu, W.; Hua, X.; Xu, C.; Lai, J.; Yuan, M.; Chen, B. Verification of concrete nonlinear creep mechanism based on meso-damage mechanics. *Constr. Build. Mater.* **2021**, *276*, 122205. [CrossRef]
18. Chen, Q.; Huang, W.; Zhang, C.; Niu, X.; Xu, X.; Deng, X.; Wang, G.; Wu, S. Experimental and analytical study of damaged reinforced concrete frame retrofit by fan-shaped lead-infilled viscoelastic dampers. *J. Build. Eng.* **2024**, *95*, 110197. [CrossRef]
19. Zhang, Y.; Pan, J.; Sun, X.; Feng, J.; Sheng, D.; Wang, H.; Zhou, X.; He, Y.; Diao, M.; Zhan, Q. Simulation of thermal stress and control measures for rock-filled concrete dam in high-altitude and cold regions. *Eng. Struct.* **2021**, *230*, 111721. [CrossRef]
20. Zhang, P.; Xiong, H.; Wang, R. Measurements and numerical simulations for cast temperature field and early-age thermal stress in zero blocks of high-strength concrete box girders. *Adv. Mech. Eng.* **2022**, *14*, 1–14. [CrossRef]
21. Luo, D.; Zhang, Z.; Li, B. The effects of shear deformation in non-rectangular steel reinforced concrete structural walls. *J. Constr. Steel Res.* **2020**, *169*, 106043. [CrossRef]
22. Wu, F.; Sheng, W.; Zhang, G.; Li, H.; Ren, Y.; Zhang, K.; Wang, C.; Sun, T. Research on the deformation and settlement characteristics of tunnel lining structures under repeated loads. *Structures* **2024**, *63*, 106366. [CrossRef]

23. Saito, S.; Maki, T.; Tsuchiya, S.; Watanabe, T. Strain based damage indicators for finite element analysis of reinforced concrete structures. *Structures* **2023**, *58*, 105473. [CrossRef]
24. Liu, S.; Zhang, Y.; Zhou, B.; Yang, K. Numerical simulation of autoclaved aerated concrete masonry wall subjected to close-in explosion and the structural damage assessment. *Sci. Rep.* **2024**, *14*, 3928. [CrossRef] [PubMed]
25. Wang, Z. Research on concrete temperature damage model. Ph.D. Thesis, Hohai University, Nanjing, China, 2001.
26. Wang, X. Application of concrete damage theory in simulation analysis of hydraulic structures. Ph.D. Thesis, Hohai University, Nanjing, China, 2004.
27. Tang, C.; Zhu, W. *Numerical Test of Concrete Damage and Fracture*; Science Press: Beijing, China, 2003.
28. Zhang, J.; Yu, M.; Chu, X. Coupled hygro-thermo-mechanical peridynamic model for concrete fracture analysis at high temperature. *Comput. Part. Mech.* **2023**, *11*, 1661–1680. [CrossRef]
29. Zhang, H.L.; Davie, C.T. A numerical investigation of the influence of pore pressures and thermally induced stresses for spalling of concrete exposed to elevated temperatures. *Fire Saf. J.* **2013**, *59*, 102–110. [CrossRef]
30. Zhang, K.; Huang, T.; Yang, J.; Zhang, F.; Lei, J.; Yuan, Q. Influences of cyclic air temperature and solar radiation on the long-term deterioration behaviour of double-block ballastless track. *Constr. Build. Mater.* **2024**, *440*, 137431. [CrossRef]
31. Wang, H.; Li, L.; Du, X. A multiaxial thermo-mechanical damage model for concrete including Load Induced Thermal Strain (LITS). *Structures* **2024**, *67*, 106975. [CrossRef]
32. Seng, B.; Magniont, C.; Gallego, S.; Lorente, S. Behavior of a hemp-based concrete wall under dynamic thermal and hygric solicitations. *Energy Build.* **2021**, *232*, 110669. [CrossRef]
33. Zhu, J.; Wang, C.; Yang, Y.; Wang, Y. Hygro-thermal–mechanical coupling analysis for early shrinkage of cast in situ concrete slabs of composite beams: Theory and experiment. *Constr. Build. Mater.* **2023**, *372*, 130774. [CrossRef]
34. Shi, J.; Liu, B.; Wu, X.; Qin, J.; Jiang, J.; He, Z. Evolution of mechanical properties and permeability of concrete during steam curing process. *J. Build. Eng.* **2020**, *32*, 101796. [CrossRef]
35. Zhu, B.F. *Thermal Stresses and Temperature Control of Mass Concrete*; China Water and Power Press: Beijing, China, 2012.
36. *JGJ55-2011*; Specification for Mix Proportion Design of Ordinary Concrete. Standardization Administration of China: Beijing, China, 2011.
37. Lima, H.S.; Tsallis, C. Ising chain: Thermal conductivity and first-principle validation of Fourier's law. *Phys. A Stat. Mech. Its Appl.* **2023**, *628*, 129161. [CrossRef]
38. Fairbairn, E.M.; Silvoso, M.M.; Ribeiro, F.L.; Toledo-Filho, R.D. Determining the adiabatic temperature rise of concrete by inverse analysis: Case study of a spillway gate pier. *Eur. J. Environ. Civ. Eng.* **2017**, *21*, 272–288. [CrossRef]

buildings

MDPI

Article

Safety Status Prediction Model of Transmission Tower Based on Improved Coati Optimization-Based Support Vector Machine

Xinxi Gong [1,2], Yaozhong Zhu [3], Yanhai Wang [1,2,*], Enyang Li [1,2], Yuhao Zhang [1,2] and Zilong Zhang [1,2]

1 College of Electricity and New Energy, China Three Gorges University, Yichang 443002, China; 202208580121056@ctgu.edu.cn (X.G.); 202208580121103@ctgu.edu.cn (E.L.); 202208580121335@ctgu.edu.cn (Y.Z.); 202208580121338@ctgu.edu.cn (Z.Z.)
2 Hubei Provincial Research Center for Transmission Line Engineering Technology, China Three Gorges University, Yichang 443002, China
3 State Grid Corporation of China Chongqing Electric Power Company Ultra High Voltage Branch, Chongqing 400039, China; garu1423@126.com
* Correspondence: yanhaiwang@ctgu.edu.cn

Abstract: Natural calamities have historically impacted operational mountainous power transmission towers, including high winds and ice accumulation, which can result in pole damage or diminished load-bearing capability, compromising their structural integrity. Consequently, developing a safety state prediction model for transmission towers may efficiently monitor and evaluate potential risks, providing early warnings of structural dangers and diminishing the likelihood of bending or collapse incidents. This paper presents a safety state prediction model for transmission towers utilizing improved coati optimization-based SVM (ICOA-SVM). Initially, we optimize the coati optimization algorithm (COA) through inverse refraction learning and Levy flight strategy. Subsequently, we employ the improved coati optimization algorithm (ICOA) to refine the penalty parameters and kernel function of the support vector machine (SVM), thereby developing the safety state prediction model for the transmission tower. A finite element model is created to simulate the dynamic reaction of the transmission tower under varying wind angles and loads; ultimately, wind speed, wind angle, and ice cover thickness are utilized as inputs to the model, with the safe condition of the transmission tower being the output. The predictive outcomes indicate that the proposed ICOA-SVM model exhibits rapid convergence and high prediction accuracy, with a 62.5% reduction in root mean square error, a 59.6% decrease in average relative error, and a 75.0% decline in average absolute error compared to the conventional support vector machine. This work establishes a scientific foundation for the safety monitoring and maintenance of transmission towers, effectively identifying possible dangers and substantially decreasing the likelihood of accidents.

Keywords: structural safety; stability of buildings transmission towers; coati optimization algorithm; refractive inverse learning; Levy flight; support vector machines

1. Introduction

The rising demand for electricity in society makes transmission towers crucial for electric energy transmission, which is directly linked to the safe operational capacity essential for societal development and stability [1]. Due to their operational characteristics and geographical placement, transmission towers, particularly in mountainous regions, endure diverse static and dynamic loads over extended durations, including strong winds, heavy rainfall, and other intricate environmental conditions. This exposure renders them susceptible to the loosening of bolted connections, corrosion of joints, and deformation of structural components [2,3]. These variables will result in a deterioration of the load-bearing capacity of transmission towers, potentially leading to severe power incidents such as tower deformation or collapse [4]. Consequently, evaluating safety conditions for transmission towers in dynamic mountainous regions is essential.

Regarding the safety assessment of transmission towers, Chen et al. proposed a framework utilizing 3D point clouds and risk indices to accurately identify damage risks in flooding conditions and assist in grid maintenance [5]; Fu et al. examined the gust response coefficients of transmission towers during typhoons, discovering that these coefficients surpassed current standards, and recommended enhancing the gust response coefficient to ensure transmission tower safety [6]; Li et al. assessed the impact of wind fatigue on the brittleness of transmission towers subjected to multiple catastrophes using a probabilistic analytical framework. The findings indicated a significant reduction in structural capacity over time, offering guidance for the design and maintenance of transmission towers [7]; Zhu et al. proposed a method to evaluate the vulnerability of transmission towers in a downburst wind, and through the simulated wind field and parameterized analysis, they obtained the limit carrying capacity and the most unfavorable location, and showed that the lower storm winds are more destructive than the atmospheric boundary layer winds [8]; Zhu et al. proposed a wind vulnerability analysis method for transmission towers based on the wind direction, and determined the optimal layout and probability of failure, which provided a basis for the design and risk assessment of the transmission towers [9]; Kwon et al. analyzed the degradation of structural performance caused by corrosion of the 765 kV transmission towers in coastal areas, and a vulnerability assessment method was proposed and an ultimate collapse surface was created to assess its structural safety under different wind speeds and wind angles of attack [10]; Meng et al. analyzed the response of microtopography to transmission towers under the influence of typhoons through virtual wind tunnel simulation, and found that the microtopography significantly increased the local wind speed and the risk of collapse, which provided an essential reference for the assessment of the wind resistance of transmission lines under complex terrain [11]; Gao et al. examined the dynamic reaction of ice removal on wineglass-shaped transmission towers and determined that the central V-shaped section is more susceptible to collapse. They recommend reinforcing the critical linking component to enhance overall stability [12]; Mukherjee et al. examined the reliability of anchor foundations for transmission towers subjected to fluctuating wind loads and determined that soil cohesion significantly influences the resistance to uprooting capacity through a multivariate adaptive regression spline model. They underscored the necessity of accounting for variability to enhance structural stability and safety [13].

In terms of monitoring and warning of transmission towers, Shen et al. proposed a technique using clustered dynamic mode decomposition and adaptive block compression sensing to accurately monitor the looseness of transmission towers and safeguard their structural safety through MIMU signal analysis [14]; Min et al. designed a high-sensitivity fiber Bragg grating smart rod to achieve high-precision strain monitoring and temperature compensation for large-span transmission towers, which provides a reliable solution for transmission tower safety monitoring [15]; Huang et al. explored the effect of trains passing on the vibration response of transmission towers, and found that it reduces the specific modal frequency, which may lead to misjudgment of health monitoring, and thus optimized the vibration monitoring device to improve the safety [16]; Jin et al. predicted the foundation settlement displacement of transmission towers in the Salt Lake area by using the multi-temporal phase interferometric synthetic aperture radar, which provides a good solution for the transmission network's ground settlement disaster monitoring [17]; Tian et al. developed a fiber Bragg grating flexible sensor with temperature self-compensation to realize automatic early warning of uneven settlement and horizontal sliding of tower base by real-time monitoring of three-dimensional deformation field of the tower base of transmission towers [18]; Lu et al. researched the factors related to the deformation of the ground surface of the transmission line in the mining area, and constructed a risk assessment and early warning system to effectively predict hidden dangers and improve transmission line safety [19]; Yao et al. proposed a multi-parameter localization method based on a hybrid distributed fiber optic sensing system, which significantly improves the localization accuracy and efficiency of transmission towers, and provides reliable support for online monitoring and early warning of high-voltage transmission lines [20]; He et al. conducted millimeter surface displacement of transmission lines with synthetic aperture radar (SAR) interferometry millimeter-level monitoring of surface displacements along transmission lines in the Loess Plateau by synthetic aperture radar interferometry, which provides critical data support for the early warning of geologic hazards in intelligent grids [21]; Zhang et al. designed a fiber Bragg grating-based on-line monitoring system for the transmission of tower strain, which achieves accurate early warning of transmission tower failures and provides effective support for condition maintenance and disaster prevention [22].

Despite the advancements in the security evaluation of transmission towers presented in the studies above, significant deficiencies still need to be addressed. These investigations predominantly depend on several data sources, including meteorological, geographic, and grid information, and exhibit a relative need for more intelligent algorithms. Although offering essential context for the evaluation, this dependence leads to inadequate model precision to accurately represent the actual risk in extreme weather scenarios or intricate geographic settings. Furthermore, the velocity and precision of forecasts are constrained, hindering the capacity for real-time monitoring and swift response. To enhance the efficacy of assessment models, it is imperative to implement sophisticated, intelligent algorithms, such as machine learning and deep learning, which can manage extensive, multi-dimensional data, discern potential nonlinear relationships, and consequently augment models' predictive capacity and adaptability. Future studies should concentrate on incorporating intelligent technologies to enhance the invention and optimization of assessment models, thereby attaining a superior level of safety management.

The support vector machine (SVM) is a prevalent supervised learning method mainly employed for multi-class classification tasks [23,24]. It efficiently streamlines the computational procedure by selecting suitable hyperplanes and employing the concept of high and low-dimensional transformations, showcasing considerable benefits in examining nonlinear, high-dimensional, and small-sample issues [25,26]. Nonetheless, the conventional SVM exhibits prolonged training durations and elevated computational complexity when confronted with extensive datasets, and there exists variability in the values of the penalty parameter c and the kernel function parameter g [27]. This paper proposes an improved coati optimization algorithm (ICOA) to optimize the SVM's penalty and kernel function parameters, effectively mitigating overfitting and underfitting, addressing slow computation speed, and improving the accuracy of the model's classification results.

Currently, limited research has employed intelligent algorithms to assess the safety of transmission towers, and the prevailing evaluation methods for these towers exhibit significant subjectivity, as well as low accuracy and efficiency in their assessments. This research offers a transmission tower safety status model utilizing the improved coati optimization algorithm (ICOA) to enhance the SVM. The coati optimization algorithm (COA) is initially enhanced using back-refraction learning and the Levy flight strategy. The SVM's penalty parameters and kernel functions are optimized utilizing the improved coati optimization algorithm (ICOA) to develop the transmission tower safety condition prediction model. Secondly, a finite element model of the transmission tower is constructed using ABAQUS 2020 software, and the dynamic response of the tower is simulated under various wind angles and loads. The simulated dataset and the ICOA-SVM method are employed to develop a predictive model for the safety condition of transmission towers, utilizing wind angle, wind speed, and ice thickness as inputs and the safety status of the transmission tower as the output. This not only enhances the maintenance and management of the transmission network but also mitigates potential safety risks and bolsters the reliability of the overall power supply, which holds significant practical significance for ensuring the safety of societal electricity use.

2. Coati Optimization Algorithm and Improvements

2.1. Coati Optimization Algorithm

The coati optimization algorithm is a novel meta-heuristic intelligent optimization technique introduced by Mohammad et al. in 2023. It is inspired by coati behaviors, specifically their methods of attacking and hunting iguanas while evading predators, to address optimization problems [28].

2.1.1. Population Initialization

The following equation can represent the original members of the coati population:

$$x_{i,j} = lb_j + r \cdot (ub_j - lb_j), i = 1,2,\ldots,N, j = 1,2,\ldots,m \tag{1}$$

where $x_{i,j}$ represents the jth dimensional position of the ith coati within the search space, r is a random actual number in the interval [0, 1], ub_j and lb_j signify the upper and lower bounds of the search for the optimal solution in the jth dimension, N indicates the number of coati populations, and m denotes the number of dimensions.

2.1.2. Hunting and Attack Strategies for Iguanas

Coatis will hunt iguanas by ascending trees, and the following equation can represent the coati's position in the tree:

$$\begin{aligned} & x_{i,j}^{P1} = x_{i,j} + r \cdot (Iguana_j - I \cdot x_{i,j}), \\ & \text{for } i = 1,2,\ldots,\left\lfloor \frac{N}{2} \right\rfloor \text{ and } j = 1,2,\ldots,m \end{aligned} \tag{2}$$

The following equation can articulate the coati's position on the ground after the iguana's landing:

$$\begin{aligned} & Iguana_j^G = lb_j + r \cdot (ub_j - lb_j), j = 1,2,\ldots,m, \\ & x_{i,j}^{P1} = \begin{cases} x_{i,j} + r \cdot (Iguana_j^G - I \cdot x_{i.j}), F_{Iguana^G} < F_i, \\ x_{i,j} + r \cdot (x_{i,j} - Iguana_j^G), else, \end{cases} \\ & \text{for } i = \left\lfloor \frac{N}{2} \right\rfloor + 1, \left\lfloor \frac{N}{2} \right\rfloor + 2,\ldots,N \text{ and } j = 1,2,\ldots,m \end{aligned} \tag{3}$$

If the coati's new position enhances the goal function's value, it is adopted; otherwise, it remains unchanged. This procedure is referred to as the greedy law and is articulated as follows:

$$x_{i,j}^{P1} = \begin{cases} x_{i,j}^{P1}, F_i^{P1} < F_i, \\ x_{i,j}, else. \end{cases} \tag{4}$$

where $x_{i,j}^{P1}$ signifies the new position of the ith coati in the jth dimension during the initial stage, r is a stochastic actual number within the interval [0, 1], $Iguana_j$ denotes the jth dimensional position of the iguana within the search space, representing the position of the optimal member of the population, and I is an integer randomly chosen from the set $\{1, 2\}$. $Iguana_j^G$ represents the jth dimensional position of the randomly produced iguana on the ground, F_i signifies the objective function of the ith coati value, and F_{Iguana^G} indicates the objective function value of the iguana on the ground.

2.1.3. Escape from Predators

When a coati confronts a predator and must evade, the following equation can represent the unpredictable destination of the coati's escape:

$$\begin{aligned} & lb_j^{local} = \frac{lb_j}{t}, ub_j^{local} = \frac{ub_j}{t}, \text{where } t = 1,2,\ldots,T. \\ & x_{i,j}^{P2} = x_{i,j} + (1-2r) \cdot (lb_j^{local} + r \cdot (ub_j^{local} - lb_j^{local})), \\ & i = 1,2,\ldots,N, j = 1,2,\ldots,m, \end{aligned} \tag{5}$$

where lb_j^{local} and ub_j^{local} represent the local lower and upper bounds of the jth dimension, respectively; t signifies the current iteration count, while T indicates the maximum iteration limit. $x_{i,j}^{P2}$ denotes the new position of the ith coati in the jth dimension during the second stage, with all other parameters defined as previously stated.

2.2. Improvement Strategies

2.2.1. Refractive Reverse Learning Initialization Population

The quality of the initialized population is pivotal in the metaheuristic algorithm, significantly influencing both the convergence speed and the correctness of the final solution. The conventional COA employs a random method for population initialization, leading to a non-uniform distribution of solution individuals within the search space, thereby impairing its global search capability and optimization efficiency. In this paper, we enhance the population initialization process by employing the refractive inverse learning strategy to augment the algorithm's performance by expanding its search range, as illustrated in Figure 1 below.

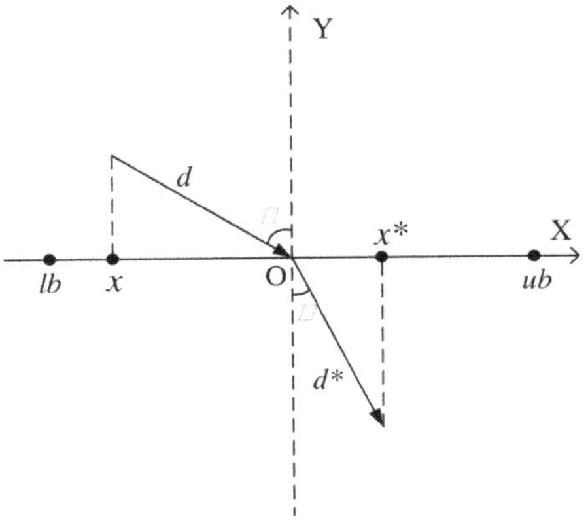

Figure 1. Schematic diagram of refractive inverse learning principle.

In Figure 1, $[lb, ub]$ on the X-axis signifies the search interval, the Y-axis represents the standard line, O denotes the midpoint of the search interval, α and β indicate the angles of incidence and refraction. d and d^* correspond to the lengths of the incident and refracted rays, while x and x^* represent the current and optimal solutions after refractive inversion learning, respectively. The refractive index n is ascertained from the geometric relationship illustrated in the picture.

$$n = \frac{\sin\alpha}{\cos\beta} = \frac{d^*((lb + ub)/2 - x)}{d(x^* - (lb + ub)/2)} \tag{6}$$

Let $k = l/l^*$, included in the equation above and extended to the multidimensional space of the algorithm, produce the refractive inverse solution $x_{i,j}^*$:

$$x_{i,j}^* = \frac{lb_j + ub_j}{2} + \frac{lb_j + ub_j}{2k} - \frac{x_{i,j}}{k} \tag{7}$$

2.2.2. Introduction of Levy Flight Strategy

In the development phase of COA, the coati modifies its location during the search process according to the current individual optimum, resulting in the algorithm prematurely converging to a local optimum solution, hence constraining its efficacy in global search. This research introduces a Levy flight technique to enhance the position update process and address this constraint.

The Levy flight approach produces randomized step lengths and broadens the search region, potentially enhancing the diversity of coati populations. The enhanced method for updating coati locations is as follows:

$$\sigma = \left(\frac{\Gamma(1+\beta)\sin\left(\frac{\pi\beta}{2}\right)}{\Gamma(\frac{1+\beta}{2})\beta \cdot 2^{\frac{\beta-1}{2}}} \right) \quad (8)$$

$$Levy(\beta) = 0.01 \cdot u \cdot \frac{r_5}{|v|^{\frac{1}{\beta}}} \quad (9)$$

Γ is the usual Gamma function, β is a random variable in the interval [0, 2], r_5 is a random variable in the interval [0, 1], and u and v follow the normal distributions $u \sim N(0, \sigma_2)$ and $v \sim N(0, 1)$, respectively.

The revised account of the coati's location after the application of the Levy flying technique is as follows:

$$x_{i,j}^{P2} = Levy(\beta) \cdot x_{i,j} + (1-2r) \cdot (lb_j^{local} + r \cdot (ub_j^{local} - lb_j^{local})) \quad (10)$$

The Levy flight strategy was introduced to the development phase of COA to enhance its global search ability and mitigate premature convergence. By generating long-distance jumps in the solution space, Levy flight enables a more diverse population distribution, improving the algorithm's capacity to escape local optima and explore the global optimum effectively.

3. Benchmark Function Testing

3.1. Baseline Function and Parameterization

Six benchmark test functions are employed for performance evaluation to assess the optimization search capability of the improved coati optimization algorithm (ICOA), with the fundamental characteristics of these functions presented in Table 1. To enhance the reliability of the test results, f_1~f_3 are selected as high-dimensional single-peak test functions, typically possessing a singular global optimal solution aimed at evaluating the algorithm's global search capability; f_4~f_6 are designated as high-dimensional multiple-peak test functions, necessitating superior local search proficiency from the algorithm, as these functions contain numerous local optimal solutions. Furthermore, six swarm optimization algorithms were selected: Grey Wolf Optimizer (GWO), Whale Optimization Algorithm (WOA), Northern Goshawk Optimization (NGO), Dung Beetle Optimizer (DBO), COA, and ICOA. The results of benchmark function tests for these algorithms were compared.

Table 1. Benchmark function.

Function	Dimension	Search Domain	Optimum
$f_1 = \sum_{i=1}^{n} x_i^2$	30	$[-100, 100]$	0
$f_2 = \sum_{i=1}^{n-1} \left[100(x_{i+1} - x_i^2)^2 + (x_i - 1)^2\right]$	30	$[-10, 10]$	0
$f_3 = \sum_{i=1}^{n} x_i^2 + \left(\sum_{i=1}^{n} \frac{1}{2} i x_i\right)^2 + \left(\sum_{i=1}^{n} \frac{1}{2} i x_i\right)^4$	30	$[-5, 10]$	0
$f_4 = -20\exp\left(-0.2\sqrt{\frac{1}{n}\sum_{i=1}^{n} x_i^2}\right) - \exp\left(\frac{1}{n}\sum_{i=1}^{n}\cos(2\pi x_i)\right) + 20 + e$	30	$[-32.768, 32.768]$	0
$f_5 = \frac{1}{4000}\sum_{i=1}^{n} x_i^2 - \prod_{i=1}^{n}\cos\left(\frac{x_i}{\sqrt{i}}\right) + 1$	30	$[-600, 600]$	0
$f_6 = 10n + \sum_{i=1}^{n}\left[x_i^2 - 10\cos(2\pi x_i)\right]$	30	$[-5.12, 5.12]$	0

To ensure the optimal performance of the coati optimization algorithm (COA), a systematic approach was adopted for parameter selection, combining initial parameter settings, sensitivity analysis, and empirical testing. The population size and maximum iterations were initially set to 30 and 500, respectively, based on values commonly used in the optimization literature. A sensitivity analysis was then conducted to evaluate the impact of these parameters on convergence speed and solution accuracy. For instance, population size was varied from 10 to 100 in increments of 10, and the results indicate that a population size of 30 provided a balance between computational efficiency and optimization quality. Additionally, parameters related to the exploration–exploitation trade-off and the Levy flight scaling factor were fine-tuned using benchmark functions f_1 to f_6. Through multiple experimental runs, these parameters were adjusted to enhance the algorithm's global search capabilities while maintaining robust performance in local search. The finalized parameters not only ensured the effectiveness of COA but also established a fair baseline for comparing its enhanced version, ICOA, with other algorithms.

3.2. Analysis of Optimization Results

The ICOA was employed to address six test functions and compared with five other algorithms. The optimal solution, mean, and standard deviation were calculated, and the statistical results are presented in Table 2.

Table 2. Validation in the effectiveness of the ICOA.

Function	Category	GWO	WOA	NGO	DBO	COA	ICOA
$f_1(x)$	Optimum	7.71×10^{-29}	8.01×10^{-89}	1.39×10^{-91}	1.55×10^{-165}	0	0
	Mean	1.37×10^{-27}	1.62×10^{-77}	1.07×10^{-87}	6.47×10^{-115}	0	0
	Standard Deviation	1.54×10^{-27}	3.90×10^{-77}	1.44×10^{-87}	2.05×10^{-114}	0	0
$f_2(x)$	Optimum	2.61×10^{1}	2.69×10^{1}	2.55×10^{1}	2.55×10^{1}	0	0
	Mean	2.69×10^{1}	2.78×10^{1}	2.59×10^{1}	2.58×10^{1}	2.36×10^{-4}	0
	Standard Deviation	8.26×10^{-1}	5.28×10^{-1}	2.58×10^{-1}	3.63×10^{-1}	7.44×10^{-4}	0
$f_3(x)$	Optimum	1.20×10^{-9}	3.88×10^{2}	5.38×10^{-10}	1.03×10^{-76}	0	0
	Mean	4.16×10^{-8}	5.32×10^{2}	2.80×10^{-8}	1.07×10^{-29}	0	0
	Standard Deviation	8.61×10^{-8}	1.10×10^{2}	8.08×10^{-8}	3.34×10^{-29}	0	0
$f_4(x)$	Optimum	8.57×10^{-14}	4.44×10^{-16}	4.00×10^{-15}	4.44×10^{-16}	4.41×10^{-16}	0
	Mean	9.92×10^{-14}	4.71×10^{-15}	6.13×10^{-15}	4.44×10^{-16}	4.41×10^{-16}	0
	Standard Deviation	1.28×10^{-14}	2.80×10^{-15}	1.83×10^{-15}	0	0	0
$f_5(x)$	Optimum	0	0	0	0	0	0
	Mean	3.83×10^{-3}	1.06×10^{-2}	0	0	0	0
	Standard Deviation	6.27×10^{-3}	3.36×10^{-2}	0	0	0	0
$f_6(x)$	Optimum	5.68×10^{-14}	0	0	0	0	0
	Mean	1.28	5.68×10^{-15}	0	2.99×10^{-1}	0	0
	Standard Deviation	2.37	1.80×10^{-14}	0	9.46×10^{-1}	0	0

The data analysis in Table 2 reveals that the ICOA method exhibits superior optimization search performance relative to other algorithms. In evaluating single-peak functions, the ICOA has superior accuracy in obtaining the optimal solution compared to other population optimization algorithms, indicating its exceptional global search capability. In the case of the multi-peak test functions f_5 and f_6, the performance disparity between ICOA and COA, both prior to and following the enhancement, is not pronounced; nonetheless, when juxtaposed with the other four algorithms that exhibit a more rapid convergence rate,

ICOA retains a competitive edge. Regarding the multi-peak function f_4, only ICOA can achieve the theoretical optimal solution, indicating its robust local search capability.

Figure 2 depicts the convergence curves of each algorithm on six benchmark test functions to intuitively illustrate the superiority of ICOA over alternative algorithms regarding optimization accuracy and convergence speed. The curves indicate that, with the same test function, the convergence speed and accuracy of ICOA are significantly superior to those of other algorithms. Furthermore, ICOA demonstrates a more rapid iteration speed under identical convergence accuracy conditions. Whether applied to single-peak or multi-peak benchmark test functions, ICOA exhibits clear advantages in the optimization process.

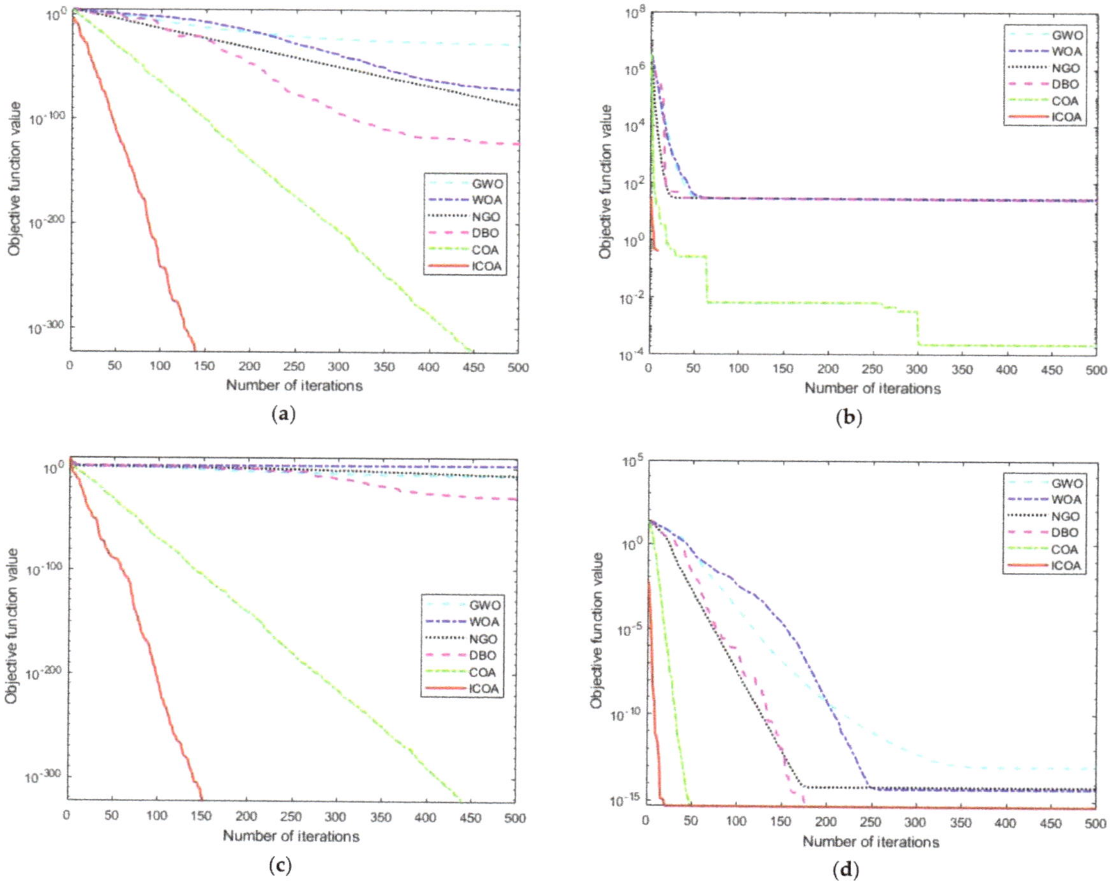

Figure 2. *Cont.*

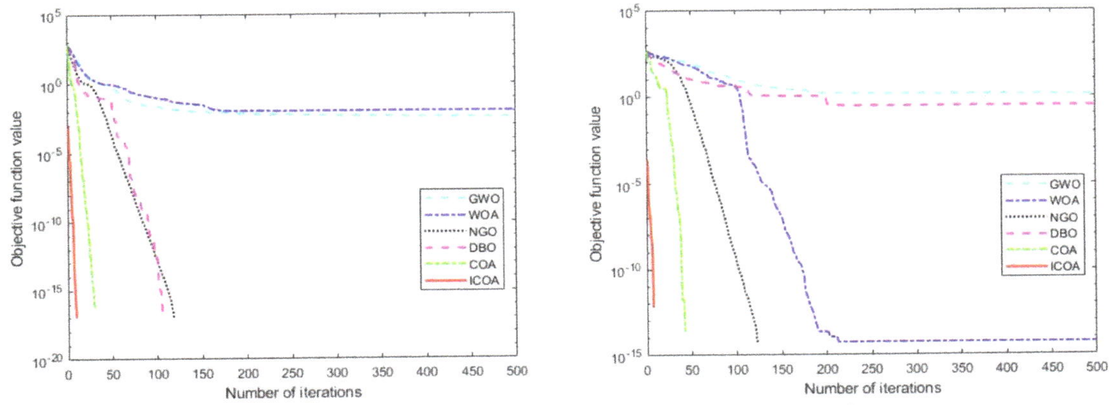

(e) (f)

Figure 2. (**a**) f_1 test function; (**b**) f_2 test function; (**c**) f_3 test function; (**d**) f_4 test function; (**e**) f_5 test function; (**f**) f_6 test function.

4. ICOA Optimized SVM Model

SVM is a supervised learning model extensively used in classification and regression analysis. Its fundamental strength is its capacity to address both linear and nonlinear data classification challenges and to identify an ideal hyperplane during the training phase, which proficiently segregates distinct data classes.

This study employs the improved ICOA to optimize the two critical parameters of SVM: the penalty factor c and the kernel function parameter g, hence enhancing SVM performance. The ICOA-SVM predicts the safety condition of transmission towers utilizing ICOA-SVM. A transmission tower safety status prediction model utilizing ICOA-SVM is built, and the samples are trained to obtain optimal classification performance. The precise sequence of the ICOA-SVM method for forecasting the safety condition of transmission towers is illustrated in Figure 3:

(1) Acquire sample data and preprocess them into two segments: training and test sets.
(2) Develop the SVM and set the appropriate parameters.
(3) Initialization of COA parameters to enhance the stochastic initialization process of the coati population with a refractive reverse learning technique.
(4) Upon concluding the cooperative phases of iguana predation and dispersal, the Levy flight strategy is implemented to revise the coati's position, finalizing the COA optimization and achieving the ICOA model.
(5) The ICOA model updates the coati's optimal location, and the revised fitness value is compared to the established optimal fitness value. The global optimal solution is achieved upon reaching a predetermined convergence accuracy or a maximum iteration limit.
(6) Utilize the optimal solution of ICOA acquired in the preceding step as the parameters of SVM (c, g), thereby constructing the ICOA-SVM classification model. Proceed to execute the model until it achieves the specified convergence accuracy or attains the maximum iteration limit, concluding the model training.
(7) Execute the trained ICOA-SVM classification model to assess the safety status of transmission towers.

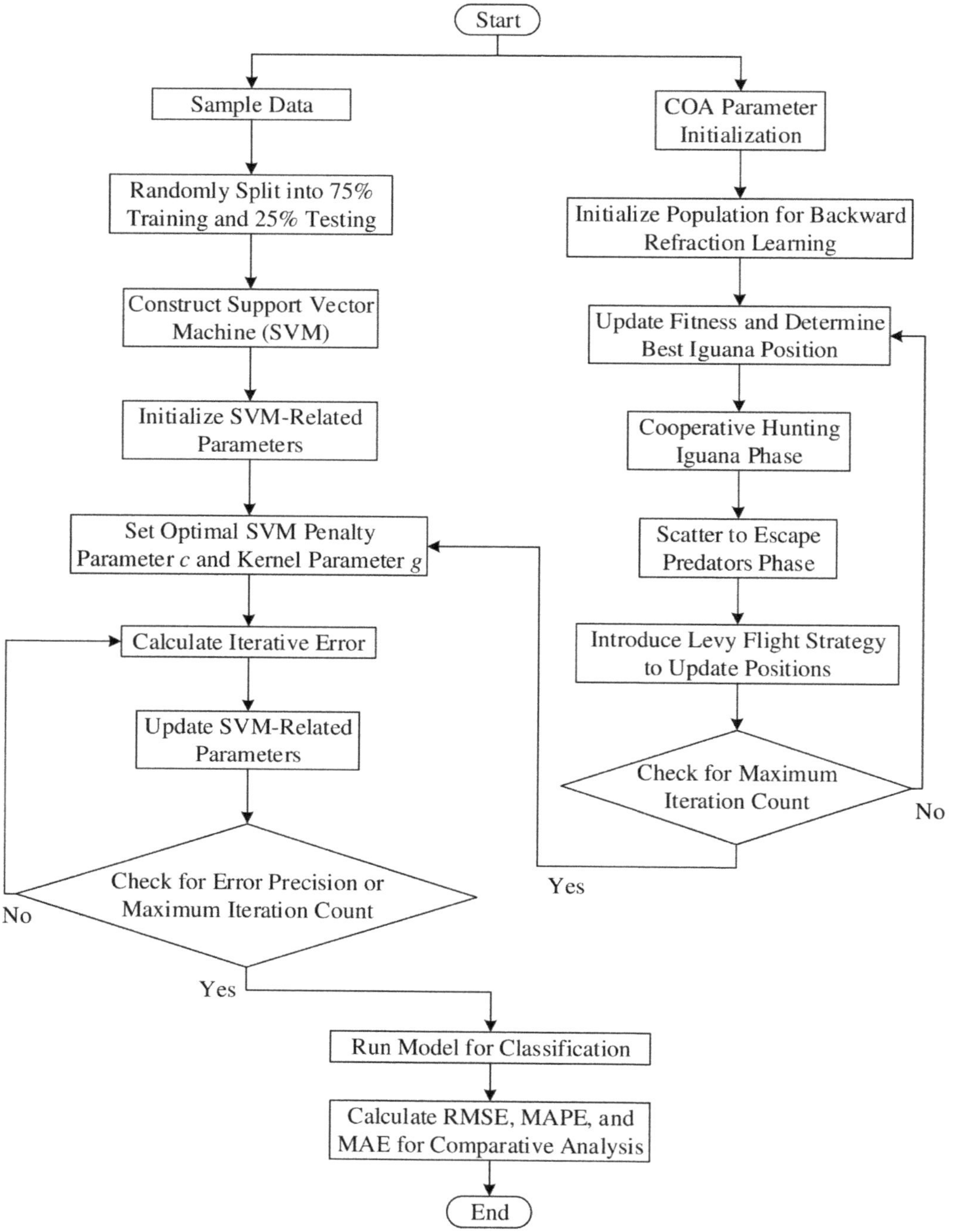

Figure 3. ICAO-SVM model flowchart.

5. Example Analysis and Predicted Results

To assess the efficacy of the ICOA-SVM prediction model in optimizing practical problem solutions, a finite element model of transmission towers was developed in ABAQUS. The ICOA-SVM model was employed to classify and predict the safety status of the transmission towers by applying ice and wind loads while also varying the wind direction angle.

5.1. Modeling

This study examines a 500 kV transmission line utilizing a straight-line tower, namely the SG2-42 model. The tower's height is 68.35 m, with a nominal height of 42 m, and the distances between the front and rear gears are 421 m and 368 m, respectively. The primary material of the transmission towers is Q420 and Q345 steel, while the auxiliary components consist of Q235 steel. Based on the features of its rod load-bearing capacity, a finite element model is established using the B31 beam unit. The material characteristics of angle steel are presented in Table 3.

Table 3. Angle steel material parameters.

Model	Poisson's Ratio	Density/(kg·m^{-3})	Young's Modulus/N·mm^{-2}	Yield Stress/MPa
Q420	0.3	7,850,000	206,000	420
Q345	0.3	7,850,000	206,000	345
Q235	0.3	7,850,000	206,000	235

Figure 4 illustrates the ABAQUS finite element model of the transmission tower.

Figure 4. Finite element model diagram of transmission tower.

5.2. Calculation and Application of Loads

Both tower–wire separation and coupling models require substantial modeling and computing effort, particularly when assessing wind speeds, wind angles, and other contributing factors. Similar static loading is employed for the finite element analysis of transmission towers to enhance efficiency and obtain a sufficiently big dataset.

Alongside the primary wind load, the transmission tower and the ground conductor must also account for their self-weight and the horizontal stress due to ice accumulation. The base of the transmission tower is established as a fixed constraint, disregarding the mutual vibrational effects between the tower and the conductor line. The horizontal wind load and the vertical load of the conductor are directly applied at the suspension point. At the same time, the tower structure is still subjected to wind load, followed by static analysis.

The self-weight load of the conductor–earth line is determined using the formula

$$G_1 = n \cdot q \cdot g \cdot L_v \tag{11}$$

The weight load of the ice is computed as

$$G_2 = 27.728n \cdot b(d+b) \cdot L_v \tag{12}$$

The associated horizontal wind loads on the conductors and ground are computed as follows:

$$W_X = \alpha \cdot W_0 \cdot \mu_z \cdot \mu_{sc} \cdot \beta_c \cdot (d+2b) \cdot L_p \cdot B \cdot \sin^2\theta \tag{13}$$

where n represents the number of sub conductors per phase; q denotes the unit mass per unit length of the overhead line; α signifies the unevenness coefficient of wind pressure; W_0 represents the standard value of the base wind pressure, while μ_z is the height-dependent coefficient of wind pressure. μ_{sc} represents the coefficient of the geometry of the conductive line. β_c is the correction coefficient for the wind load on the conductive line; d represents the outer diameter of the conductive line, whereas b denotes the thickness of the overlying ice. L_p denotes the horizontal gear distance of the tower; L_v signifies the vertical gear distance of the tower; B denotes the wind load amplification coefficient due to underlying ice; θ represents the angle between the wind direction and the ground wire direction.

5.3. Simulation Conditions and Results

The transmission tower is intended for a wind speed of 30 m/s and an ice thickness of 10 mm. A wind speed ranging from 0 to 30 m/s is applied to the transmission tower, with increments of 2 m/s. Concurrently, the ice thickness is increased by 2 mm. Wind angles of 30°, 45°, 60°, and 90° are utilized, resulting in a total of 384 distinct wind and ice load conditions for simulation calculations. To facilitate comparison and analysis, the following four representative conditions are shown as examples of finite element simulation stress and displacement clouds, as seen in Figure 5a–d.

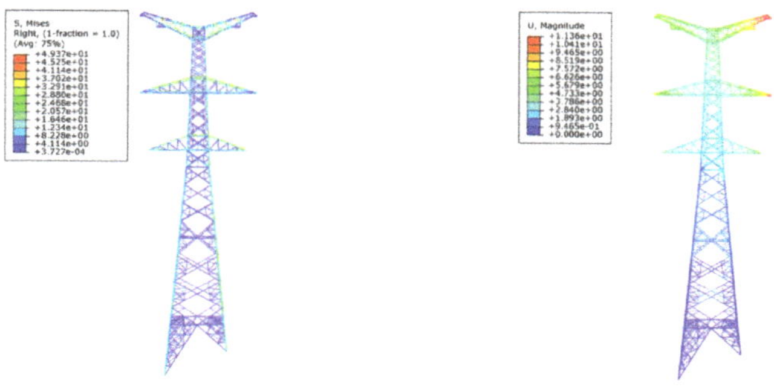

(a)

Figure 5. *Cont.*

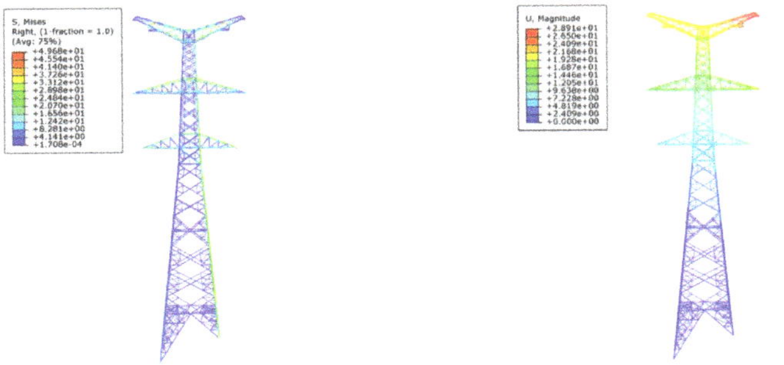

(**b**)

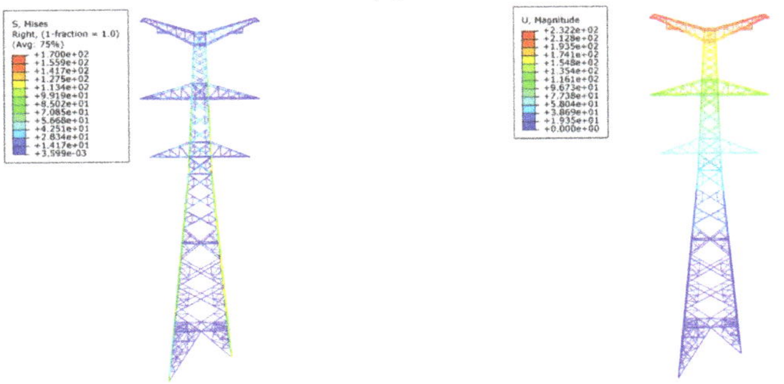

(**c**)

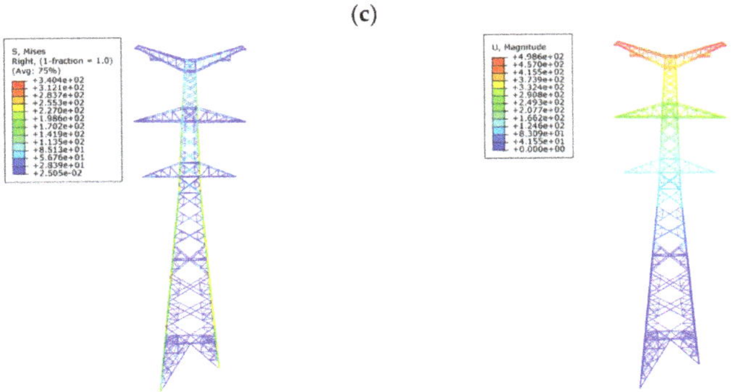

(**d**)

Figure 5. (**a**) Stress and displacement cloud diagram for a wind direction angle of 30°, wind speed of 10 m/s, and an ice thickness of 0 mm; (**b**) Stress and displacement cloud diagram for a wind direction angle of 90°, wind speed of 10 m/s, and an ice thickness of 0 mm; (**c**) Stress and displacement cloud diagram for a wind direction angle of 90°, wind speed of 30 m/s, and an ice thickness of 0 mm; (**d**) Stress and displacement cloud diagram for a wind direction angle of 90°, wind speed of 30 m/s, and an ice thickness of 0 mm.

Figure 5a,b illustrate that at a wind speed of 10 m/s, with no overlying ice, altering the wind angle does not significantly affect the maximum stress of the tower; however, the maximum displacement rises from 11.37 mm to 28.91 mm. Figure 5b,c indicate that, with a wind angle of 90 degrees, the maximum stress escalates from 49.68 MPa to 170 MPa, while the maximum displacement increases from 28.91 mm to 232.2 mm. Figure 5c,d demonstrate that the tower's maximum stress and maximum displacement rise when only the ice thickness is modified. The alteration in wind angle primarily influences the stress distribution of the tower's main material and the displacement magnitude of the tower head, whilst variations in wind speed and ice thickness impact the total stress magnitude and displacement size of the tower body.

5.4. Judging Criteria and Dataset Creation

To better illustrate the key factors influencing the safety status of transmission towers, Table 4 summarizes the primary parameters considered in this study, along with their specific impacts on structural safety. These parameters were incorporated into the finite element modeling and dataset creation processes to ensure a comprehensive representation of real-world conditions. By systematically accounting for these influencing factors, this study aims to build a robust dataset that captures the variability and complexity of real-world scenarios.

Table 4. Influencing parameters and their impact on transmission tower safety.

Parameter	Description	Impact on Safety
Wind Speed	Range: 0–30 m/s, with 2 m/s increments	Higher wind speeds significantly increase stress and displacement across the tower structure.
Wind Direction Angle	Angles of 30°, 45°, 60°, and 90°	Affects stress distribution and displacement magnitudes; maximum displacement observed at 90°.
Ice Thickness	Range: 0–10 mm, with 2 mm increments	Greater ice thickness adds weight and increases stress and displacement, particularly in vertical rods.
Operating Condition	Categories: Safe (1), Safer (2), Basic Safety (3), Unsafe (4)	The combination of wind speed, angle, and ice thickness determines the safety classification of the tower.
Tower Geometry	Tower height: 68.35 m; rod configuration	Structural dimensions influence the distribution of forces and the tower's load-carrying capacity.
Material Properties	Steel grades (Q420, Q345, Q235) with specific yield stress limits	Variations in material strength affect the tower's ability to withstand external loads.

Each parameter plays a critical role in determining the structural performance of transmission towers. For instance, wind speed significantly impacts the overall stress and displacement levels, with higher wind speeds leading to exponential increases in structural strain. Wind direction angles, such as 90°, cause asymmetrical stress distribution, often leading to maximum displacement at specific tower sections. Ice thickness, on the other hand, adds substantial weight to the structure, disproportionately affecting vertical rods and increasing both stress and displacement. These environmental factors interact with the tower's material properties and geometry, determining its load-bearing capacity and overall safety classification under various conditions. The operating condition classification system, ranging from "Safe" to "Unsafe", provides a standardized framework for quantifying safety risks. This classification is derived from a combination of key parameters, ensuring a holistic evaluation of structural stability. Additionally, the geometry and material properties of the tower define its resilience to these external forces. High-strength steel grades, such as Q420, enhance the tower's ability to withstand severe conditions, while specific geometrical configurations help optimize force distribution and minimize structural deformations. The integration of these parameters into the finite element model ensures that the ICOA-SVM prediction framework is trained and evaluated on a dataset that accurately reflects real-world operational conditions. This comprehensive approach enhances the robustness of the proposed model and underscores its applicability to diverse environmental scenarios. Future work could further refine the parameter ranges and explore additional influencing factors, such as long-term material degradation and dynamic interactions between accessories and the tower structure, to improve the model's accuracy and generalizability.

This study evaluates the safety state of transmission towers under various operational settings by analyzing their operational status and safety level through a risk assessment methodology. This risk evaluation graphically quantifies and delineates the safety level of transmission towers under various loading circumstances, offering a scientific foundation for routine operation and maintenance. Numerous experts and researchers, both domestically and internationally, have researched the assessment of transmission tower structural system failures [29,30]. The transmission tower construction is often deemed to have failed if it reaches one of the following three conditions:

(1) The structure transforms into a mechanism: the criterion for judgement is that, when some members fail, the total stiffness matrix of the remaining members exhibits singularity, resulting in the non-convergence of the ABAQUS calculation program;
(2) Structural deformation surpasses the allowable threshold: It is generally accepted that the maximum displacement response limit of the transmission tower is situated at its apex, and if the displacement at the top of the tower exceeds a specified value, it is deemed indicative of structural damage;
(3) The structure cannot endure additional loads or the initial occurrence of bearing capacity decline, mostly evidenced by excessive stress ratios in the tower's main components. This results in damage and is subsequently classified as structural failure.

Integrating the three points above and referencing the pertinent literature [31,32] and regulations [33,34], the criteria for assessing the safety state of transmission towers, as delineated in Table 5, categorize the operational state into four classifications: safe, safer, basic safety, and unsafe. Here, ξ represents the component stress ratio (the ratio of the maximum stress position to its yield stress), l denotes the tower top displacement, and h signifies the tower height.

Table 5. Criteria for judging the structural failure of transmission.

Category	Safe	Safer	Basic Safety	Unsafe
Component Stress Ratio	$\xi < 0.5$	$0.5 \leq \xi \leq 0.75$	$0.75 \leq \xi \leq 1.0$	$\xi > 1.0$
Top Tower Displacement	$l < \mathrm{h}/1000$	$\mathrm{h}/1000 \leq l \leq 2\,\mathrm{h}/1000$	$2\,\mathrm{h}/1000 \leq l \leq 3\,\mathrm{h}/1000$	$l > 3\,\mathrm{h}/1000$

According to the above transmission tower safety status assessment standard, the simulation data are classified. The tower status is divided into four categories: safe, relatively safe, basically safe, and unsafe, and the numbers 1, 2, 3, and 4 are used for abbreviation (the most unfavorable state is taken when the operating status categories are inconsistent). Then, 384 sets of data obtained by simulation are selected for experiments, and 75% of the sample data are randomly divided as the training set for the training of the transmission tower safety status prediction model. The remaining 25% of the sample data are used as the test set to test the accuracy of the model. The transmission tower safety status dataset is shown in Table 6.

Table 6. Transmission tower safety status dataset.

Wind Direction Angle/(°)	Wind Speed/(m/s)	Ice Thickness/(mm)	Operating Condition
30	10	0	1
30	30	10	2
45	30	0	2
45	20	10	2
60	30	0	3
60	30	10	4
90	10	0	1
90	20	0	2
90	20	10	3
90	30	10	4
...	...	...	...

5.5. Forecast Results and Analysis

To ascertain that the SVM model enhanced by the improved coati optimization algorithm (ICOA-SVM) possesses superior efficacy in predicting the outcomes of the safety state classification of transmission towers, a comparative analysis is conducted against the conventional SVM model and the unimproved COA-SVM model. The findings of the comparison are presented in Figures 6–8.

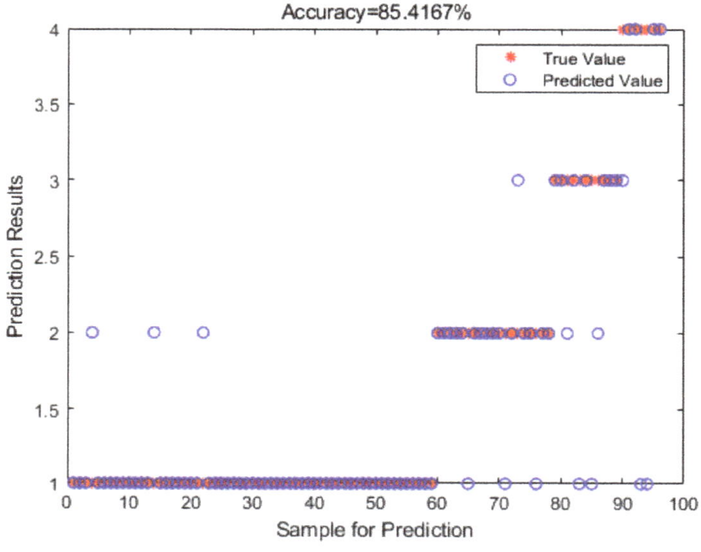

Figure 6. SVM prediction results.

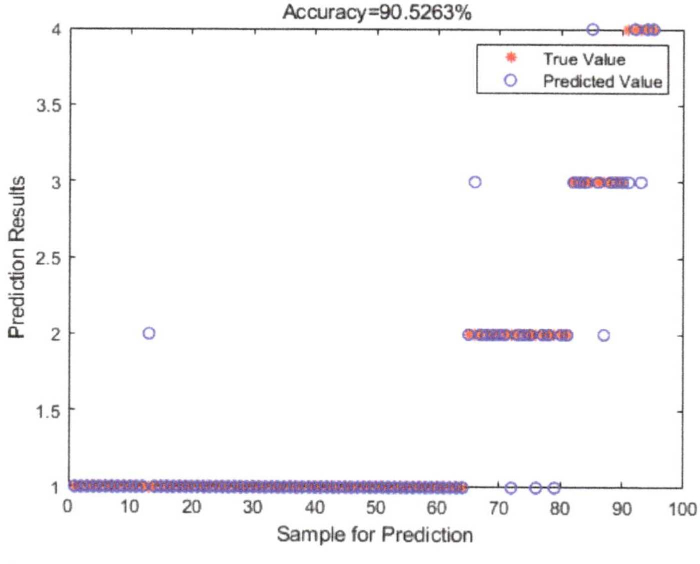

Figure 7. COA-SVM prediction results.

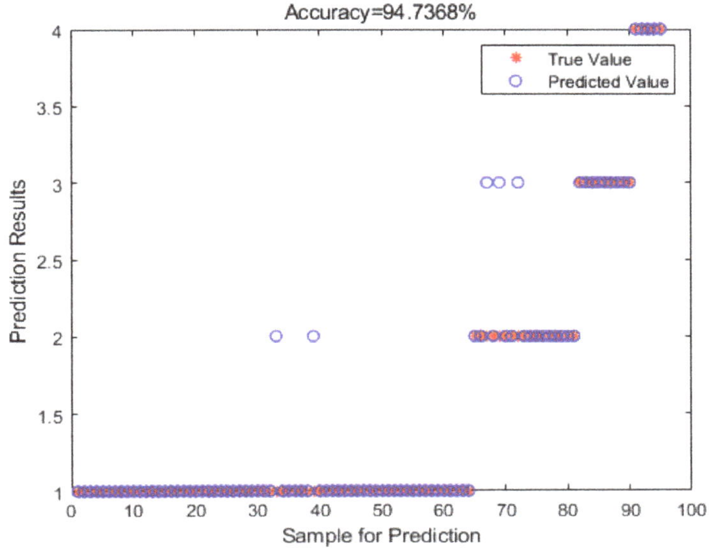

Figure 8. ICOA-SVM prediction results.

The ICOA-SVM model demonstrates superior predictive performance relative to the other two models when analyzing the same transmission tower safety state data, exhibiting minimal error with only five erroneous datasets and an accuracy of 94.7368%. In contrast, the COA-SVM model presents nine erroneous datasets and an accuracy of 90.5263%. The unoptimized SVM model shows 14 erroneous datasets and an accuracy of 85.4167%, indicating that the traditional SVM model is less effective in predicting the classification results when it is not optimized.

To effectively demonstrate that the ICOA-SVM model exhibits superior accuracy in forecasting the outcomes of safety state classification for transmission towers, we assess the predictive performance of the three models by computing the root mean square error (RMSE), the mean absolute percentage error (MAPE), and the mean absolute error (MAE). Utilize mean absolute error (MAE) to evaluate the predictive efficacy of the three models. The equations for the three error calculation formulas above are shown in Equations (14)–(16), and the computational results are displayed in Table 7.

$$RMSE = \sqrt{\frac{1}{n}\sum_{i=1}^{n}(y_i - y'_i)^2} \tag{14}$$

$$MAPE = \frac{1}{n}\sum_{i=1}^{n}\left|\frac{y'_i - y_i}{y_i}\right| \tag{15}$$

$$MAE = \frac{1}{n}\left(\sum_{i=1}^{n}|y_i - y'_i|\right) \tag{16}$$

where n represents the quantity of predicted values, y_i signifies the true value, and y_i' indicates the forecasted value.

To further ensure the robustness and generalizability of the ICOA-SVM model, a 5-fold cross-validation strategy was applied during the training phase. This approach divides the dataset into five equal parts, where four parts are used for training and the remaining part for validation in each iteration. The process is repeated five times, ensuring that each subset serves as a validation set exactly once. This strategy minimizes the risk of overfitting and provides a comprehensive evaluation of the model's performance across varying data splits. The cross-validation results demonstrate consistent improvements across all metrics (RMSE, MAPE, MAE), further validating the reliability of the proposed framework.

Table 7. Comparison results of the errors of the models.

Model \ Error	RMSE	MAPE	MAE
SVM	0.6124	0.0911	0.2083
COA-SVM	0.3078	0.0439	0.0947
ICOA-SVM	0.2294	0.0368	0.0521
RF	0.3415	0.0562	0.1017
COA-RF	0.3124	0.0497	0.0873
ICOA-RF	0.2953	0.0472	0.0828
NN	0.2843	0.0417	0.0895
COA-NN	0.2598	0.0389	0.0782
ICOA-NN	0.2476	0.0375	0.0729
GB	0.2567	0.0382	0.0728
COA-GB	0.2435	0.0361	0.0683
ICOA-GB	0.2347	0.0354	0.0651

Table 7 indicates that the SVM model exhibits the highest error values across all metrics, with an RMSE of 0.6124, a MAPE of 0.0911, and an MAE of 0.2083, highlighting its relatively poor predictive performance compared to other models. While random forests (RFs), neural networks (NNs), and gradient boosting (GB) show improved accuracy, their performance remains inferior to that of the ICOA-SVM model. The reduction in errors (RMSE, MAPE, MAE) in the ICOA-SVM model can be attributed to the efficient parameter optimization achieved by the improved coati optimization algorithm (ICOA). Specifically, the inverse refraction learning and Levy flight strategy in ICOA facilitate a more effective exploration of the solution space, preventing the algorithm from becoming trapped in local optima and ensuring optimal SVM parameters, such as the penalty parameter c and the kernel parameter g.

Notably, ICOA-SVM achieves the best results among all models, with an RMSE of 0.2294, a MAPE of 0.0368, and an MAE of 0.0521, demonstrating significant improvements of 62.5%, 59.6%, and 75.0%, respectively, over the baseline SVM model. This reduction stems from the precise tuning of SVM parameters, which allows the model to capture complex nonlinear relationships in the data more effectively than other models. Even when RFs, NNs, and GB are optimized using COA and ICOA, their errors remain slightly higher than those of ICOA-SVM, underscoring the superior suitability of SVM for this task when combined with ICOA. Additionally, the ICOA framework enhances predictive performance in RF, NN, and GB models by refining their parameter settings, yet the flexibility and structure of SVM seem to align better with the optimization capabilities of ICOA. These findings highlight the effectiveness of ICOA in enhancing predictive accuracy and the robustness of the ICOA-SVM framework for forecasting the safety status of transmission towers. Together, these results provide a reliable reference for the secure operation and proactive maintenance of critical infrastructure.

5.6. Discussion of Analytical Results

From the above results, several critical insights can be analyzed regarding the performance and applicability of the proposed ICOA-SVM model. First, the results indicate that the integration of the improved coati optimization algorithm (ICOA) with the support vector machine (SVM) significantly enhances prediction accuracy compared to the conventional SVM and COA-SVM models. As shown in Table 7, the ICOA-SVM model achieved the lowest RMSE (0.2294), MAPE (0.0368), and MAE (0.0521), representing reductions of 62.5%, 59.6%, and 75.0%, respectively, when compared to the traditional SVM. These substantial reductions in error metrics highlight the effectiveness of the ICOA in optimizing SVM parameters, specifically the penalty parameter c and kernel parameter g. The use of inverse refraction learning and the Levy flight strategy within the ICOA framework ensured better exploration of the solution space, thereby mitigating issues of local optima that often hinder conventional optimization methods.

Second, the finite element modeling (FEM) results provide valuable insights into the structural response of the transmission towers under different environmental conditions. For instance, the stress and displacement patterns in Figure 5a–d reveal that the wind angle significantly influences the stress distribution and displacement magnitude, while wind speed and ice thickness directly impact the overall stress and displacement levels. These observations validate the importance of incorporating diverse environmental factors, such as wind speed, wind angle, and ice thickness, into the training data for the predictive model. By simulating a comprehensive range of conditions (384 cases), the FEM results provided a robust dataset for training and evaluating the ICOA-SVM model, ensuring its applicability to real-world scenarios.

Third, the classification results from the ICOA-SVM model demonstrate its robustness and adaptability. As seen in the safety state predictions, the model accurately classified the transmission tower's safety status across varying conditions with an accuracy of 94.74%, outperforming both COA-SVM (90.53%) and traditional SVM (85.42%). This consistency across different environmental scenarios underscores the model's ability to generalize well, making it a reliable tool for structural risk assessment. The ability to handle diverse input conditions ensures that the ICOA-SVM model can provide early warnings for potentially unsafe states, supporting proactive maintenance and reducing the likelihood of structural failures.

Lastly, the practical implications of this study are significant. The proposed methodology not only improves prediction accuracy but also offers a scalable approach for assessing the safety of transmission towers in mountainous or harsh environments. By combining advanced simulation techniques, such as FEM, with intelligent optimization algorithms, this study establishes a framework that can be extended to other structural systems requiring safety assessments. The results demonstrate that the ICOA-SVM model is a reliable and efficient tool for power transmission infrastructure management, enabling early detection of structural risks and reducing maintenance costs while ensuring operational reliability.

6. Research Implications

The findings of this study have both theoretical and practical implications. Theoretically, the proposed ICOA-SVM model extends the application of intelligent optimization algorithms to structural safety prediction. By integrating advanced techniques like inverse refraction learning and Levy flight strategies, this study improves the predictive capabilities of SVMs for nonlinear, high-dimensional, and small-sample problems. These advancements contribute to the broader field of machine learning and optimization, offering a novel approach for addressing safety-critical tasks in infrastructure management.

Practically, this study provides a reliable and efficient methodology for assessing the safety of transmission towers under complex environmental conditions. The ICOA-SVM model demonstrated superior performance in accurately predicting the safety states of transmission towers across a range of wind speeds, wind angles, and ice thicknesses. This predictive capability enables timely detection of unsafe conditions, supporting proactive maintenance strategies and reducing the risk of structural failures. Furthermore, the methodology is scalable and can be adapted to other types of infrastructure, making it valuable for improving the reliability and resilience of power transmission systems in diverse geographic and environmental contexts.

The findings also highlight the model's adaptability to varying geographic conditions. While this study primarily focuses on transmission towers in mountainous regions, the ICOA-SVM framework is flexible and can be applied to non-mountainous areas by incorporating location-specific environmental factors such as wind speed, wind angles, and ice loads into the dataset. This adaptability ensures the model's practical utility for diverse transmission tower types, supporting its scalability across a wide range of geographic and environmental contexts. Future studies can further validate its effectiveness by incorporating real-world data from non-mountainous terrains to ensure robust generalization.

These implications underscore the significance of combining finite element simulations with intelligent optimization techniques, paving the way for future studies to explore similar approaches in other fields, such as civil engineering, transportation, and renewable energy systems.

7. Conclusions

This paper addresses safety issues that may arise in mountain transmission towers due to complex natural conditions, including rod damage and reduced bearing capacity. We propose a prediction model for the safety status of transmission towers utilizing the ICOA-SVM, enabling the forecasting of their safety state. The following conclusions are drawn:

(1) During the finite element simulation, the force condition of the tower structure was assessed by varying the wind angle, wind speed, and ice thickness. The alteration of wind angle primarily influences the stress distribution of the tower's main material and the displacement of the tower head. However, variations in wind speed and ice thickness impact both the force and displacement of the entire tower structure.

(2) This paper addresses the prolonged training time and computational complexity of traditional SVM methods by employing reverse refraction learning and incorporating the Levy flight strategy to enhance the COA. The resulting ICOA demonstrates superior convergence speed and optimization capability compared to alternative algorithms.

(3) Propose an ICOA-SVM for predicting the safety state of transmission towers, demonstrating superior prediction accuracy compared to the SVM. The RMSE value of the ICOA-SVM model decreased by 62.5%, the MAPE value decreased by 59.6%, and the MAE value decreased by 75.0%. This advancement can provide a scientific foundation for the safe operation of the electric power system by enhancing the accuracy and efficiency of the assessment.

Although the proposed ICOA-SVM model demonstrates significant advancements in predicting the safety status of transmission towers, certain limitations need to be acknowledged. Firstly, the dataset used in this study is derived from finite element simulations under predefined conditions. While these simulations provide a controlled and comprehensive framework, they may not fully capture the complexity and variability in real-world scenarios, such as unexpected structural defects, extreme environmental conditions, dynamic interactions with hanging accessories like insulators and conductors, or long-term material degradation. Secondly, the computational complexity of the ICOA framework, although improved compared to traditional methods, remains a limitation, particularly for large-scale applications or real-time monitoring systems. This complexity may pose challenges when scaling the model to process large datasets or operate in environments requiring rapid decision-making. Thirdly, this study focuses specifically on transmission towers, which limits the model's generalizability to other types of infrastructure. Similar predictive frameworks could potentially be adapted and validated for other structures, such as bridges, pipelines, or wind turbines. These limitations emphasize the need for further research to enhance the model's computational efficiency, expand its applicability to diverse infrastructure systems, and validate its performance using real-world datasets to ensure its broader applicability and practical feasibility.

Future research should aim to address these limitations by expanding the dataset to include real-world measurements under more diverse environmental conditions, thereby improving the model's robustness and enhancing its generalizability to complex, real-world scenarios. Efforts should also focus on improving the computational efficiency of the ICOA framework, potentially through further refinements in algorithmic design or the integration of hybrid approaches. Such advancements could enable the deployment of the ICOA-SVM model in real-time monitoring systems, providing timely and accurate safety assessments for critical infrastructure. Additionally, to better capture the nonlinear relationship between wind speed and direction, future studies will include finer granularity in wind direction data beyond the current discrete angles (30°, 45°, 60°, 90°). This refinement will ensure a more comprehensive understanding of how varying wind directions impact the safety and stability of transmission towers.

Additionally, the application of the ICOA-SVM model should also be extended to other structural systems, such as bridges, wind turbines, or pipelines, to demonstrate its adaptability and effectiveness across a broader range of contexts. Incorporating factors like material aging, long-term fatigue effects, and dynamic loading conditions into the model could further enhance its predictive accuracy and comprehensiveness. These improvements would not only refine the current methodology but also establish a more scalable, adaptive, and practical framework for managing infrastructure safety in diverse and challenging environments. By addressing these areas, the proposed approach has the potential to significantly contribute to the fields of structural engineering and intelligent infrastructure management.

Author Contributions: Conceptualization, X.G. and Y.W.; methodology, X.G.; software, X.G. and E.L.; validation, X.G., Y.Z. (Yaozhong Zhu) and Z.Z.; formal analysis, X.G., Y.Z. (Yuhao Zhang), E.L. and Z.Z.; investigation, X.G.; resources, Y.W.; data curation, X.G.; writing—original draft preparation, X.G.; writing—review and editing, X.G., Y.Z. (Yaozhong Zhu), Y.W., E.L., Y.Z. (Yuhao Zhang) and Z.Z.; supervision, Y.W.; project administration, Y.Z. (Yaozhong Zhu); funding acquisition, Y.W. All authors have read and agreed to the published version of the manuscript.

Funding: This research was funded by the National Natural Science Foundation of China, grant number: 52079070.

Data Availability Statement: The data that support the findings of this study are available from the corresponding author upon reasonable request.

Conflicts of Interest: Author Yaozhong Zhu was employed by the company State Grid Corporation of China Chongqing Electric Power Company Ultra High Voltage Branch. The remaining authors declare that the research was conducted in the absence of any commercial or financial relationships that could be construed as a potential conflict of interest.

References

1. Shair, J.; Li, H.; Hu, J.; Xie, X. Power system stability issues, classifications and research prospects in the context of high-penetration of renewables and power electronics. *Renew. Sustain. Energy Rev.* **2021**, *145*, 111111. [CrossRef]
2. Liu, S.; Zhang, W.; Li, Q.; Yan, S.; Zhang, S.; Li, C.; Li, L. Dynamic Failure Mode Analysis for a Transmission Tower-Line System Induced by Strong Winds. *Energies* **2024**, *17*, 4679. [CrossRef]
3. Zhang, L.; Ren, Y.; Zhou, X.; Shen, G.; Tu, Z.; Yao, J. Wind-Induced Responses of Corroded Angle-Steel Transmission Tower. *Energies* **2023**, *16*, 5429. [CrossRef]
4. Zhou, Y.; Sheng, Q.; Chen, J.; Li, N.; Fu, X.; Zhou, Y. The failure mode of transmission tower foundation on the landslide under heavy rainfall: A case study on a 500-kV transmission tower foundation on the Yanzi landslide in Badong, China. *Bull. Eng. Geol. Environ.* **2022**, *81*, 125. [CrossRef]
5. Chen, M.; Chan, T.O.; Wang, X.; Luo, M.; Lin, Y.; Huang, H.; Sun, Y.; Cui, G.; Huang, Y. A risk analysis framework for transmission towers under potential pluvial flood-LiDAR survey and geometric modelling. *Int. J. Disaster Risk Reduct.* **2020**, *50*, 101862. [CrossRef]
6. Fu, X.; Du, W.-L.; Li, H.-N.; Xie, W.-P.; Xiao, K.; Lei, X. Gust Response Factor of a Transmission Tower Under Typhoon. *Int. J. Struct. Stab. Dyn.* **2021**, *21*, 2150001. [CrossRef]
7. Li, C.; Pan, H.; Tian, L.; Bi, W. Lifetime multi-hazard fragility analysis of transmission towers under earthquake and wind considering wind-induced fatigue effect. *Struct. Saf.* **2022**, *99*, 102266. [CrossRef]
8. Zhu, C.; Yang, Q.; Wang, D.; Huang, G.; Liang, S. Fragility Analysis of Transmission Towers Subjected to Downburst Winds. *Appl. Sci.* **2023**, *13*, 9167. [CrossRef]
9. Zhu, C.; Yang, Q.; Huang, G.; Zhang, X.; Wang, D. Fragility analysis and wind directionality-based failure probability evaluation of transmission tower under strong winds. *J. Wind. Eng. Ind. Aerodyn.* **2024**, *246*, 105668. [CrossRef]
10. Kwon, K.; Choi, Y.; Choi, Y.; Han, W.S.; Kim, J.H.; Kong, J.S. Extended limit-collapsed surfaces using fragility analysis of high voltage transmission towers located in coastal areas under wind load. *Adv. Struct. Eng.* **2024**, *27*, 2116–2132. [CrossRef]
11. Meng, X.; Tian, L.; Ma, R.; Zhang, L.; Liu, J.; Dong, X. Typhoon-induced failure analysis of electricity transmission tower-line system incorporating microtopography. *Eng. Fail. Anal.* **2024**, *163*, 108556. [CrossRef]
12. Gao, S.; Zeng, C.; Zhou, L.; Liu, X.; Gao, B. Numerical analysis of the dynamic effects of wine-cup shape power transmission tower-line system under ice-shedding. *Structures* **2020**, *24*, 1–12. [CrossRef]

13. Mukherjee, S.; Pramanik, R.; Babu, G.L.S. Probabilistic assessment of anchor foundation for transmission tower using multivariate adaptive regression splines. *Georisk-Assess. Manag. Risk Eng. Syst. Geohazards* **2024**, *18*, 541–557. [CrossRef]
14. Shen, L.; Yang, J. Looseness Monitoring of Transmission Tower Based on MIMU. *IEEE Sens. J.* **2024**, *24*, 1655–1664. [CrossRef]
15. Xue, M.; Zhu, C.; Li, B.; Zhao, W.; Chu, F.; Zhu, R.; Yuan, B. Design and characteristic research of smart rod based on optical fiber grating for strain monitoring of power transmission tower. *Sci. Prog.* **2023**, *106*, 00368504231171250. [CrossRef]
16. Huang, X.; Liu, Z.; Zhao, L.; Yuan, P.; Zhang, Y. Research on Vibration Analysis and Online Monitoring Technology of Transmission Tower Near Railway. *IEEE Instrum. Meas. Mag.* **2023**, *26*, 34–41. [CrossRef]
17. Jin, B.; Zeng, T.; Yang, T.; Gui, L.; Yin, K.; Guo, B.; Zhao, B.; Li, Q. The Prediction of Transmission Towers' Foundation Ground Subsidence in the Salt Lake Area Based on Multi-Temporal Interferometric Synthetic Aperture Radar and Deep Learning. *Remote Sens.* **2023**, *15*, 4805. [CrossRef]
18. Tian, C.; Ma, X.; Ma, X.; Peng, B. Optimal design of FBG flexible sensor for high-precision monitoring of three-dimensional deformation of power transmission line tower foundation. *Energy Rep.* **2023**, *9*, 46–54. [CrossRef]
19. Lu, Z.; Xu, S.; Wang, S.; Meng, X.; Hou, F. Research on risk assessment technology for transmission lines in mining affected areas. *AIP Adv.* **2024**, *14*, 105309. [CrossRef]
20. Yao, Y.; Wang, R.; Ding, H.; Tong, S.; Han, Y.; Zhao, S.; Zou, N.; Xiong, F.; Zhang, Y. A Fast and Accurate Mapping Method for an OPGW Tower Based on Hybrid Distributed Optical Fiber Sensing. *Sensors* **2024**, *24*, 5629. [CrossRef]
21. He, H.; Zhou, L.; Lee, H. Ground Displacement Variation Around Power Line Corridors on the Loess Plateau Estimated by Persistent Scatterer Interferometry. *IEEE Access* **2021**, *9*, 87908–87917. [CrossRef]
22. Zhang, L.; Ruan, J.; Du, Z.; Huang, D.; Deng, Y. Transmission line tower failure warning based on FBG strain monitoring and prediction model. *Electr. Power Syst. Res.* **2023**, *214*, 108827. [CrossRef]
23. Hu, N.; Kong, L.; Zheng, H.; Zhou, X.; Wang, J.; Tao, J.; Li, W.; Lin, J. Trend Prediction of Vibration Signals for Pumped-Storage Units Based on BA-VMD and LSTM. *Energies* **2024**, *17*, 5331. [CrossRef]
24. El-Sayed, E.I.; ElSayed, S.K.; Alsharef, M. Data-Driven Approaches for State-of-Charge Estimation in Battery Electric Vehicles Using Machine and Deep Learning Techniques. *Sustainability* **2024**, *16*, 9301. [CrossRef]
25. He, Y.-L.; Liang, J.-T.; Tian, Y.; Zhu, Q.-X. Novel Schur Decomposition Orthogonal Exponential DLPP With Mixture Distance for Fault Diagnosis. *IEEE Trans. Ind. Inform.* **2024**, *20*, 5601–5608. [CrossRef]
26. Dang, C.; Wei, P.; Beer, M. An approach to evaluation of EVD and small failure probabilities of uncertain nonlinear structures under stochastic seismic excitations. *Mech. Syst. Signal Process.* **2021**, *152*, 107468. [CrossRef]
27. Xia, Y.; Zhang, C.; Wang, C.; Liu, H.; Sang, X.; Liu, R.; Zhao, P.; An, G.; Fang, H.; Shi, M.; et al. Prediction of bending strength of glass fiber reinforced methacrylate-based pipeline UV-CIPP rehabilitation materials based on machine learning. *Tunn. Undergr. Space Technol.* **2023**, *140*, 105319. [CrossRef]
28. Dehghani, M.; Montazeri, Z.; Trojovska, E.; Trojovsky, P. Coati Optimization Algorithm: A new bio-inspired metaheuristic algorithm for solving optimization problems. *Knowl. Based Syst.* **2023**, *259*, 110011. [CrossRef]
29. Zhang, J.; Xie, Q. Failure analysis of transmission tower subjected to strong wind load. *J. Constr. Steel Res.* **2019**, *160*, 271–279. [CrossRef]
30. Dikshit, S.; Dobson, I.; Alipour, A. Cascading structural failures of towers in an electric power transmission line due to straight line winds. *Reliab. Eng. Syst. Saf.* **2024**, *250*, 110304. [CrossRef]
31. Tian, L.; Pan, H.; Ma, R.; Zhang, L.; Liu, Z. Full-scale test and numerical failure analysis of a latticed steel tubular transmission tower. *Eng. Struct.* **2020**, *208*, 109919. [CrossRef]
32. Fu, X.; Du, W.-L.; Li, H.-N.; Li, G.; Dong, Z.-Q.; Yang, L.-D. Stress state and failure path of a tension tower in a transmission line under multiple loading conditions. *Thin Walled Struct.* **2020**, *157*, 107012. [CrossRef]
33. *DL/T 5154*; Technical Code for the Design of Tower and Pole Structures of Overhead Transmission Line. China Planning Press: Beijing, China, 2012.
34. *GB 50545*; Code for Design of 110kV~750kV Overhead Transmission Line. China Planning Press: Beijing, China, 2010.

buildings

Article

Damage Characteristics of Blasting Surrounding Rock in Mountain Tunnel in Fault Fracture Zones Based on the Johnson–Holmquist-2 Model

Lizhi Cheng [1], Zhiquan Yang [1,*], Ping Zhao [2] and Fengting Li [3]

[1] Faculty of Public Safety and Emergency Management, Kunming University of Science and Technology, Kunming 650093, China
[2] Shandong Province Transportation Planning and Design Institute Group Co., Jinan 250031, China
[3] School of Civil Engineering, Shandong University, Jinan 250061, China
* Correspondence: yzq1983816@kust.edu.cn

Abstract: Blasting is a widely employed technique for tunnel construction in mountainous regions; however, it often causes damage to the surrounding rock mass, particularly in fault fracture zones, which can lead to hazards such as rockfalls and collapses. This study examines the characteristics of damage to surrounding rock due to tunnel blasting through fault fracture zones. Based on an actual tunnel blasting construction project, we conducted a finite element analysis using the JH-2 material model, taking into account the width of the fault fracture zone. Results indicate that as the width of the fault fracture zone increases, the disturbance effect of tunnel blasting on the surrounding rock becomes more pronounced. Compared to the arch bottom and arch waist of the tunnel, the tunnel vault primarily absorbs the slip deformation and compressive forces resulting from blasting disturbances in the fault fracture zone. The findings of this paper contribute a valuable methodology for analyzing the mechanical mechanisms in mountain tunnel blasting and provide essential theoretical parameters to inform the design and construction of tunnel blasting projects.

Keywords: mountain tunnels; blasting construction; JH-2 model; numerical calculations

1. Introduction

During blasting construction of mountain tunnels through fault fracture zones, differential stress–strain variations are often induced in the surrounding rock, potentially leading to damage and failure of the tunnel structure. Understanding the dynamic deformation characteristics of the surrounding rock at fault fracture zones is crucial for ensuring safe construction through these complex areas. Currently, numerical methods are predominantly employed to study the mechanical behavior in mountain tunnel construction; however, the selection of an appropriate constitutive model tailored to tunnels crossing fault fracture zones remains uncommon.

Based on typical engineering practices, the vibration characteristics of tunnel secondary lining structures under the influence of fault zones were studied through the establishment of a finite element numerical model, and a vibration velocity safety criterion with stress as the control standard was proposed [1]. Using this stress criterion and an elastic–plastic ontological model, a statistical damage model of the rock mass was developed, allowing the study of cumulative damage evolution and distribution in grouted and reinforced peripheral rock under blasting and excavation conditions [2]. Applying fracture mechanics theory, a numerical simulation analysis investigated the dynamic response and vibration attenuation of surrounding rock during blasting and excavation near fault control zones, highlighting the impact of dynamic excavation on fault stability [3]. Geological and numerical analyses showed that the laminar-structured fault fragmentation zone presents a heterogeneous state, affecting stress redistribution in tunnels and contributing to significant tunnel deformation [4]. Numerical simulations were conducted to assess surface blasting, analyzing changes in effective stress around and perpendicular to the blast hole under blasting loads [5]. Single and multiple full-face blasts were modeled by embedding a rock damage model into LS-DYNA R14.0 software through user-defined subroutines, with simulations verified against field test data [6]. The impact of millisecond delay blasting on surrounding rock damage was evaluated using acoustic testing and numerical simulation, with different detonator hole positions examined for their effects on surrounding rock integrity [7]. Cyclic blasting loads were analyzed for their effects on tunnel surrounding rock damage using a self-constructed physical model test system simulating high in situ stress conditions [8]. Additionally, the effects of millisecond delay blasting were investigated through a case study of the Chongqing North Avenue Tunnel [9]. Four scaled tunnel arch structures were fabricated and tested for anti-blast performance under internal explosions, with static load tests conducted on both undamaged and blast-damaged specimens to assess residual bearing capacities [10]. Finally, challenges in achieving smooth blasting results were addressed during the construction of the Liuzhi–Anlong Expressway Tunnel [11].

In summary, numerical calculation and analytical methods are predominantly applied to study the construction mechanics of tunnels traversing fault fracture zones. However, challenges remain regarding the selection of appropriate constitutive models and simulation techniques for blasting in these faulted areas. This research focuses on the blasting operations at the Lianfeng Tunnel entrance along the Dayong Expressway in Yunnan Province, China. Utilizing a finite element analysis program, JWL blasting theory, and the JH-2 material model, this study examines the characteristics of blast-induced damage to the surrounding rock. The findings aim to elucidate the deformation mechanisms in the surrounding rock due to tunnel blasting in fault fracture zones, offering valuable theoretical parameters and guidance for the design and execution of mountain tunnel blasting projects.

2. Numerical Calculation Models and Analysis Methods

To elucidate the dynamic deformation characteristics of blasting in the surrounding rock of a tunnel passing through a fault fracture zone, this study was conducted using the Lianfeng Tunnel Project on the Dayong Expressway in Yunnan Province, China as a case study.

2.1. Project Overview

The Lianfeng Tunnel on the Dayong Expressway is a separated double-bore tunnel with a vertical clearance of 5 m and a width clearance of 10.25 m. The tunnel alignment follows a nearly straight path with an axial direction of approximately 309°. The left bore, classified as a long tunnel, spans 10,984 m with a gradient of −2.600% and reaches a maximum depth of approximately 732 m at chainage K33 + 240. The right bore is similar in length at 10,995 m, also with a gradient of −2.600%, and reaches a maximum depth of about 748 m at chainage K33 + 190, beginning at K20 + 583 and ending at K31 + 578. The tunnel crosses the Lianfeng Fracture Zone (F12), which is the primary fracture structure in the survey area. This zone is characterized by northeast-trending faults oriented at approximately 40° to 45°, dipping between 310° and 315°, with fault surface inclinations of about 80° to 85°. The fracture exhibits clockwise torsional characteristics, affecting the strata from the Upper Aurignacian Unified Lamp Shadow Formation to the Lower Triassic Unified Feixiankuan Formation, with more significant damage to the latter. The faults encountered by the tunnel are highly destructive to the strata, impacting the surrounding rock, which is a moderately weathered gray rock with extensively developed joints. This rock mass is highly fractured with a brecciated structure and possesses extremely poor self-stabilizing capabilities. This is shown in Figure 1.

Tunnel construction employs conventional mine blasting techniques. In fault-fractured zones, the three-step construction method is utilized, with light surface blasting technology applied to strictly control over-excavation and under-excavation. The blasting advance should be determined based on the surrounding rock conditions, with excavation in weak rock masses limited to 1–2 m per blast. During tunnel blasting, the step lengths should be adjusted according to the rock conditions: When the surrounding rock at the tunnel crown is unstable, construction support must closely follow the excavation. In such cases, extending step lengths can help reduce construction interference, ensuring quality excavation, effective support, and overall construction safety. The step height is set according to geological conditions, tunnel section dimensions, and other factors, with the initial height of the upper step set at 4.5 m, adjustable as needed during excavation. When constructing the steel frame for the upper step, measures such as widening the arch foot or using foot-locking anchors should be implemented to control surrounding rock deformation and secure the initial support. Excavation of the middle and lower steps should proceed only when the sprayed concrete on the upper step reaches at least 70% of its design strength. In cases of unstable rock, the pace should be shortened, and, if necessary, the middle and lower steps can be excavated in a staggered manner, with immediate application of initial support and arch. A composite lining structure is applied in the tunnel, with initial support including measures such as Φ8 mm steel mesh, a steel frame, C25 shotcrete, Φ22 mortar anchors, and Φ25 hollow anchors.

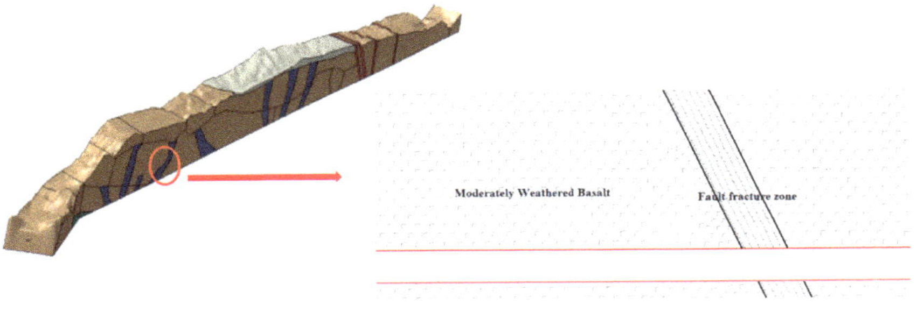

Figure 1. Schematic diagram of the distribution of fault fracture zones of Lianfeng Tunnel.

2.2. Rupture Modeling Theory

For explosion simulations, various equations of state have been developed to characterize the behavior of explosive products, with the Jones–Wilkins–Lee (JWL) equation of state widely recognized as one of the most comprehensive [12,13]. The JWL equation is frequently chosen in blasting analysis due to its effectiveness in modeling the dynamic response of high explosives, particularly in capturing pressure and volume changes during detonation. Tailored to simulate the rapid, high-pressure conditions of explosive detonation, JWL theory is instrumental in predicting the shockwave behavior immediately after detonation—essential for assessing blast impacts on surrounding structures. The parameters within the JWL equation are typically derived from empirical data specific to each explosive, allowing it to deliver validated, reliable results across a range of commonly used explosives. As it provides a more precise pressure–volume relationship for explosive gases than other general equations of state, the JWL equation is especially valuable for predicting damage and deformation in nearby structures, such as in tunnel construction.

The JWL equation, introduced in the 1960s by scientists John Paul Wierzbicki and John Wilkins, is particularly effective in modeling the strain energy changes in materials under compressive loads. It accurately characterizes the dynamic behavior of materials subjected to high pressures, elevated temperatures, and high strain rates, making it especially suitable for analyzing responses to dynamic loading scenarios, such as high-speed impacts and explosions. Due to its robustness, the JWL equation has become widely used in numerical simulations. The equation of state is expressed as follows:

$$P = Ae^{-R_1 V} + Be^{-R_2 V} + CV^{-(1+\omega)} \tag{1}$$

In the JWL equation, A, B, C, R_1, R_2, and ω are explosion constants, while v represents the relative volume v_1/v. These constants are typically determined by calibrating hydrodynamic simulation results against data from a cylindrical metal expansion test, commonly known as the Cylinder Test.

During the initial phase, the explosion products are extremely hot, especially as the gas undergoes adiabatic expansion, passing through the Chapman–Jouguet (CJ) point and entering the rapid expansion phase. At this stage, the behavior of the explosion products closely approximates that of an ideal gas. The energy contained in the explosion products during expansion can be expressed as follows:

$$e = \frac{A}{R_1}e^{-R_1 V} + \frac{B}{R_2}e^{-R_2 V} + \frac{C/\omega}{V^{\omega}} \tag{2}$$

The two formulas above share the same parameters, and the JWL equation of state, incorporating the energy term, can be derived through a system of simultaneous equations:

$$P = A\left(1 - \frac{\omega}{VR_1}\right)e^{-R_1 V} + B\left(1 - \frac{\omega}{VR_2}\right)e^{-R_2 V} + \frac{\omega e}{V^{\omega}} \tag{3}$$

The parameters in this equation, as well as the pressure and energy generated per unit volume of the explosion product in the CJ state and the detonation velocity, can be entered as input parameters in ABAQUS 2021.

2.3. Tunnel Blasting Constitutive Model

The choice of an appropriate material model is critical for the accuracy of numerical simulations. In geotechnical engineering, applying a model that closely replicates the stress–strain characteristics of actual structures is essential. Under blast loads, the surrounding rock in a tunnel typically transitions through elastic, plastic, and yielding phases before ultimately failing. Consequently, selecting a suitable constitutive model for rock damage is crucial. This study adopts the widely used Johnson–Holmquist-2 (JH-2) material model, developed by Gordon R. Johnson and Tim J. Holmquist, which introduces the Hugoniot Elastic Limit (HEL) to effectively characterize rock damage under blast impacts. The JH-2 model comprises three main components: a strength model, a damage model, and a set of corresponding polynomial equations.

Figure 2 presents the strength model of the JH-2 material constitutive model. Based on the damage factor D, the material is classified into three states: intact, damaged, and fractured. The standardized equivalent stresses corresponding to these states are depicted in the figure, with A, B, C, M, and N representing material parameters. Furthermore, T and p indicate the material's tensile and compressive strength limits, respectively.

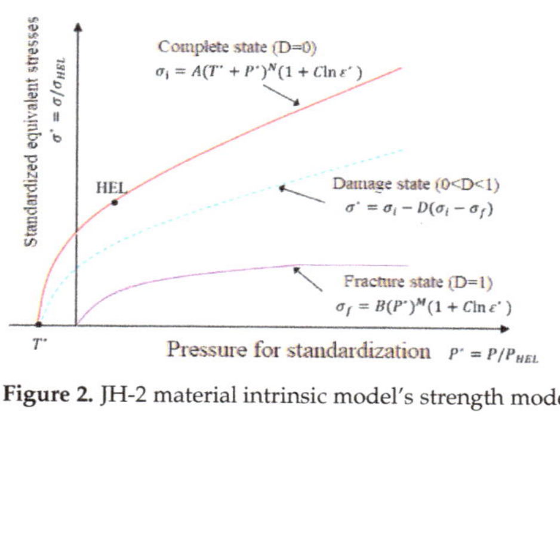

Figure 2. JH-2 material intrinsic model's strength model.

2.4. Selection of Mechanical Parameters

In this numerical simulation, the parameters for the JWL explosives are presented in Table 1, while the calculation parameters for the surrounding rock are derived from the field investigation report, with values provided in Table 2.

Table 1. Parameters of the JWL equation of state for emulsion explosives.

Density/ρ (kg·m^{-3})	Explosive Velocity/D (m·s^{-1})	A/GPa	B/GPa	R_1	R_2	ω	E_0/GPa
1600	6930	557.60	5.35	6.10	1.07	0.35	4.19

Table 2. Perimeter rock parameters.

	Density/ρ (kg·m^{-3})	Elastic Modulus/MPa	Poisson's Ratio	Cohesion/kPa	Internal Friction Angle
Moderately weathered basalt	2942	4372	0.23	214	33
Fault fracture zone	2574	273	0.28	97	25

The Johnson–Holmquist-2 model is commonly used to simulate the behavior of brittle materials under high-velocity impacts and high pressures. Determining the mechanical parameters for the JH-2 model involves a series of experimental and computational methods. The numerical calculation methods and mechanical parameters of the thesis mainly refer to the literature [14–16]. The specific rock mass parameters for the JH-2 model are shown in Table 3.

Table 3. Parameters of JH-2 constitutive model of the basalt rock mass.

ρ (kg·m^{-3})	G/MPa	A	B	C	M	N	T/MPa	HEL/GPa
2574	98.6	1.14	0.35	0.0049	0.65	0.682	43	3.3
P_{HEL}/GPa	β	D_1	D_2	K_1/GPa	K_2/GPa	K_3/GPa	σ_{Fmax}	\
2.28	1.0	0.005	0.6	43	−259	1203	0.24	\

A tunnel blasting model utilizes the Jones–Wilkins–Lee (JWL) equation of state to simulate the explosive blast process. The JWL equation effectively models the pressure exerted on the surrounding rock due to the explosion, capturing the explosive response and detonation characteristics. The detonation time and response of the explosive are influenced by the geometry of the charge, specifically the distance between the material point and the detonation point, and the velocity of the blast wave. The pressure generated during the explosive blast is described by the following equation:

$$p = A\left(1 - \frac{\omega\rho}{R_1\rho_0}\right)exp\left(-R_1\frac{\rho_0}{\rho}\right) + B\left(1 - \frac{\omega\rho}{R_2\rho_0}\right)exp\left(-R_2\frac{\rho_0}{\rho}\right) + \omega\rho E_m \quad (4)$$

where p is the blast pressure generated by the explosion; A and B are material constants of the explosive; R_{ho} is the density of the blast product; R_{ho} subscript 0 is the user-defined initial density of the explosive; ω is the Gruneisen parameter; R_1 and R_2 are dimensionless parameters; and E_m represents the initial internal energy of the explosive.

In ABAQUS/Dynamic-Explicit, the blast wave is applied to the numerical model cell through the combustion fraction.

$$F_b = min\left[1, \frac{\left(t - t_d^{mp}\right)C_d}{B_S L_e}\right] \quad (5)$$

where B_s is a constant (typically 2.5) that controls the width of the blast wave, $t_d{}^{mp}$ represents the time taken for the blast wave to reach the material point, L_e denotes the characteristic length of the cell, and C_d is the velocity of the blast wave in the explosive material. The pressure generated by the explosive is dependent on $t_d{}^{mp}$. If $t < t_d{}^{mp}$, the pressure remains zero; otherwise, the pressure is computed as the product of the pressure and the fraction of material burned, as described by the JWL equation. The time for the explosive blast wave to propagate is calculated using the following equation:

$$t_d^{mp} = min\left[t_d^N + \frac{\sqrt{\left(x^{mp} - x_d^N\right)\cdot\left(x^{mp} - x_d^N\right)}}{C_d}\right] \quad (6)$$

2.5. Numerical Calculation Models

According to the geological conditions and the actual situation of the project, the numerical simulation uses finite difference software Abaqus 2021 and establishes a three-dimensional calculation model, which intercepts two adjacent high-risk fracture zones on the mountain (480 m wide × 850 m long × 450 m high), with the widths of the two fracture zones of 67 m and 48 m, respectively, and the spacing of the zones of 73 m, and the tilt angle of 6°. In order to improve the quality of the grid, the tunnel is simplified into a circular tunnel with a diameter of 7.2 m, in which the spacing between the two tunnels is 40 m, the thickness of the lining structure is 0.7 m, and the support structure is established in accordance with the reinforcement diagram of the tunnel to support the inner and outer two ring steel frames, and different spacing of the steel frames is set for simulation. The mesh of the model is of C3D8R type. This is shown in Figure 3.

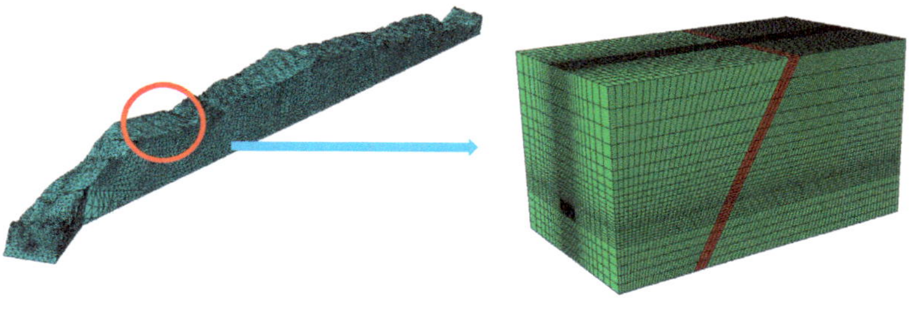

Figure 3. Numerical calculation model.

2.6. Numerical Calculation and Analysis Methods

To comprehensively simulate the damage patterns of two weathered granites under varying ground stresses during blasting, a series of detailed numerical simulations were conducted using ABAQUS finite element analysis software. This section outlines the critical steps of the numerical analysis process to ensure accurate and reliable simulation results. The key steps are as follows:

(1) Analysis Step Configuration: The numerical simulation comprises two analysis steps. In Step 1, the prestress field is imported to simulate ground stress under different peripheral pressures. Step 2 involves the blast simulation, where the duration of the analysis step is set to 4 ms, based on a review of experimental results and accuracy calculations [17].

(2) Boundary Conditions: To simulate the effect of geostress on rock blasting, static pressure is applied to the model. In addition to the pressure on the surface of the blast hole, vertical geostress is represented by static pressure in the Y direction, while horizontal geostress is represented by static pressure in the X and Z directions.

(3) Geostress Balancing: To accurately replicate the natural ground stress state, a static analysis is first performed. In this step, boundary conditions are applied to the same meshed rock model, and the resulting static stress is used to generate an ODB file. In the subsequent dynamic blast simulation, this file is used to create a prestress field. By

defining the loading steps and time points, the initial ground stresses for each rock unit are established before blasting.

(4) Contact Conditions: General contact between the components is defined using the Explicit method. Tangential and normal behaviors are specified, with friction behavior following the "penalty" formulation and a friction coefficient of 0.1. Normal contact behavior is defined as "hard contact".

(5) UMAT Subroutine: In the material module, the user material (UMAT) subroutine is invoked. The material model includes 32 attributes, and the non-independent variable function is called with eight solution-dependent state variables. The input file is generated, and material model parameters and output variables are defined by referencing the JH-2 model.

In tunnel blasting construction, analyzing the damage to surrounding rock with fault width as a variable is essential, as fault width directly affects the rock's capacity to absorb and distribute blasting energy. A wider fault zone typically contains weaker, more fractured rock, making it more susceptible to further damage, instability, and potential fracture propagation. In contrast, a narrower fault zone may better withstand blasting stresses but could also transmit shock waves more directly to adjacent areas, resulting in distinct damage patterns. Therefore, setting the fault zone's inclination angle as a fixed parameter and examining the surrounding rock damage and deformation characteristics by varying the fault width provide valuable guidance for actual project construction.

2.7. Reliability Analysis

Using the subway tunnel at Danshan Station in Qingdao as a case study, the JH-2 model was introduced as the constitutive model for the tunnel blasting simulation. The parameters of the JH-2 model specific to the local surrounding rock were obtained through experiments. Subsequently, numerical simulations and theoretical verifications were conducted to evaluate the safety distance of shotcrete based on various safety assessment criteria. The results demonstrate that this numerical calculation and analysis method can accurately assess the blasting damage characteristics of tunnel surrounding rock and serves as a valuable tool for engineering damage characterization [14].

3. Blasting Damage Characteristics of Tunnel Surrounding Rock in Fault Fracture Zones

3.1. Stress–Strain Characteristics of the Tunnel Surrounding Rock in Fault Fracture Zones

Figure 4 presents a cloud diagram illustrating the stress distribution in the surrounding rock caused by tunnel blasting across varying fault widths—specifically 10 m, 20 m, and 30 m. The figure highlights the following observations:

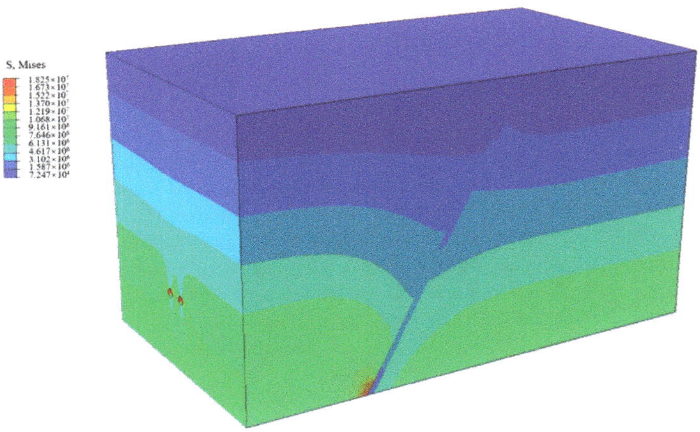

(**a**) Fault width of 10 m.

Figure 4. *Cont.*

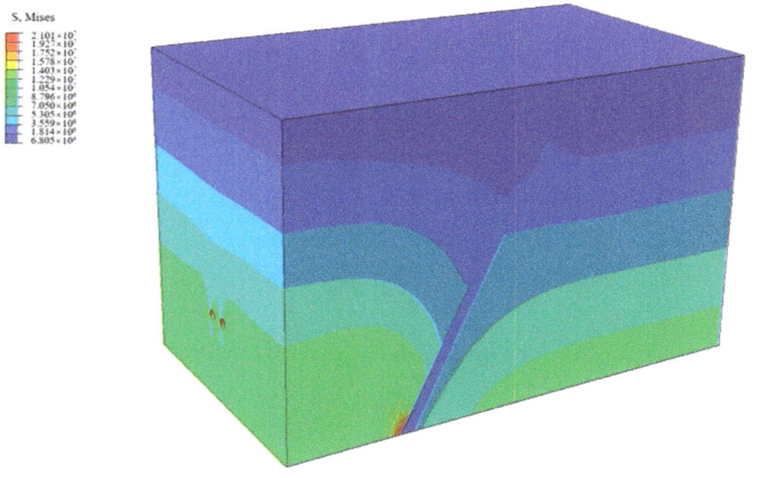

(**b**) Fault width of 20 m.

(**c**) Fault width of 30 m.

Figure 4. Stress distribution of surrounding rock induced by tunnel blasting in fracture area.

(1) Narrow Fault Widths: When the fault width is relatively small, the rock within the fault zone remains more intact, leading to stress concentration near the fault itself. This localized stress intensification can aggravate damage to the rock mass surrounding the tunnel, especially at the fault boundary where high-stress areas are likely to develop.

(2) Wide Fault Widths: In contrast, when the fault width is larger, the rock mass within the fault zone may already be highly fragmented or loosened. In such cases, the stress waves generated by blasting are absorbed or attenuated as they propagate through the fault. Consequently, stress variations within the fault zone are less significant, but the surrounding rock outside the fault can experience greater stress, particularly in the transition zone between the fault and the intact rock mass.

(3) Mechanical Properties of the Fault Zone: The mechanical characteristics of the rock within the fault zone (e.g., elasticity modulus and strength) often differ significantly from those of the surrounding intact rock. An increase in fault width results in notable

stress redistribution both inside and outside the fault zone. Wide faults tend to form large low-stress zones, while narrow faults contribute to high-stress concentration zones.

Given the concentration of stress in the surrounding rock, the tunnel support design must account for enhanced local reinforcement, particularly near the fault zone boundaries. This includes careful consideration of anchor placement, steel arches, and shotcrete application.

In conclusion, fault width plays a critical role in stress variation during tunnel blasting. Narrow faults typically lead to stress concentrations, while wide faults may induce slip within the fault zone and destabilize a broader area of the surrounding rock mass. Therefore, when tunneling through fault zones, it is essential to comprehensively evaluate fault width, surrounding rock mechanics, blasting methods, and support design to ensure the safety and stability of the construction.

Given the concentration of stress in the surrounding rock, the tunnel support design must account for enhanced local reinforcement, particularly near the fault zone boundaries. This includes careful consideration of anchor placement, steel arches, and shotcrete application. In conclusion, fault width plays a critical role in stress variation during tunnel blasting. Narrow faults typically lead to stress concentrations, while wide faults may induce slip within the fault zone and destabilize a broader area of the surrounding rock mass. Therefore, when tunneling through fault zones, it is essential to comprehensively evaluate fault width, surrounding rock mechanics, blasting methods, and support design to ensure the safety and stability of the construction.

Figure 5 presents a cloud diagram illustrating the longitudinal variation in surrounding rock strain caused by tunnel blasting across different fault widths (10 m, 20 m, and 30 m). The figure reveals several key observations:

(1) Strain Response in Narrow Faults: When the fault width is small, the fault absorbs and reflects less of the stress wave generated by blasting. As a result, stress is concentrated in the surrounding rock near the fault zone, leading to a more pronounced strain response, particularly in the transition areas on either side of the fault. This localized strain concentration can cause greater deformation in the surrounding rock post-blasting, especially at the tunnel vault and the sides of the excavation. In contrast, wider faults exhibit a different strain pattern. Due to the loose or fractured nature of the rock mass within the fault zone, the stress wave is attenuated and scattered, reducing stress transfer to the surrounding rock outside the fault. Consequently, the strain within the fault zone itself may be larger, while the surrounding rock outside the fault experiences a smaller strain response. This scenario typically results in a large plastic deformation zone within the fault, while the strain outside remains comparatively limited.

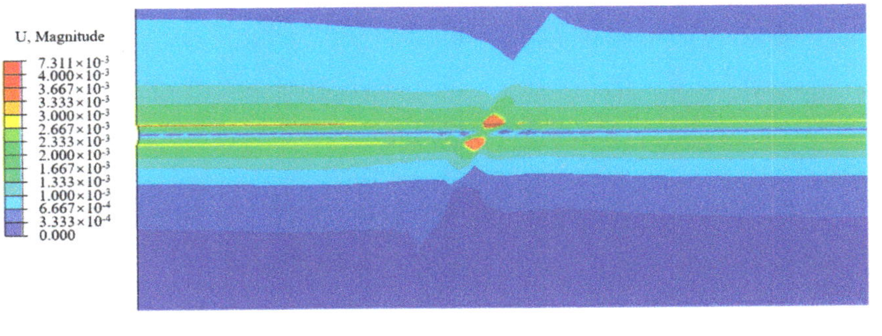

(**a**) Fault width of 10 m

Figure 5. *Cont.*

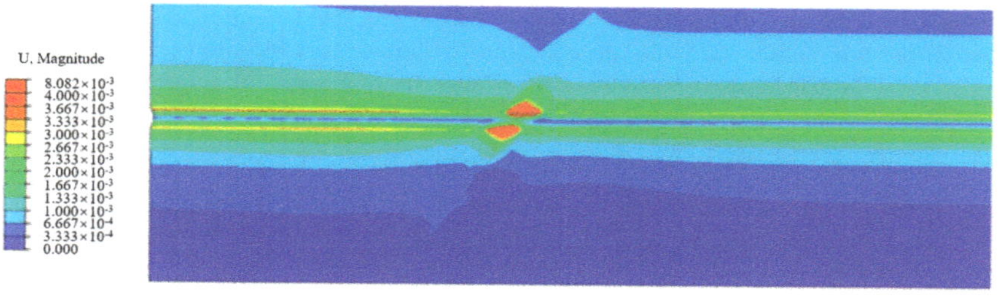

(**b**) Fault width of 20 m

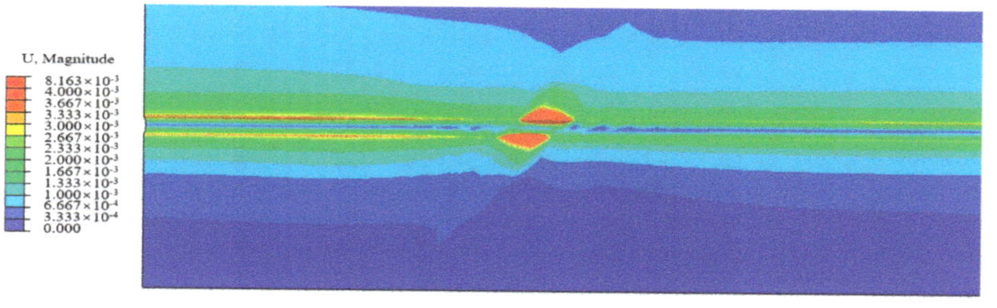

(**c**) Fault width of 20 m

Figure 5. Strain distribution of surrounding rock induced by tunnel blasting in fracture area.

(2) Deformation Characteristics of Narrow and Wide Faults: In narrow fault zones, where the rock is relatively intact, post-blast strain tends to concentrate locally around the fault, especially at the intersection of the tunnel and the fault. This strain concentration can lead to tensile or shear deformation, increasing the risk of localized damage to the surrounding rock. In contrast, within wide fault zones, the rock is generally looser or more fragmented, resulting in significant plastic deformation when the blast stress wave passes through. This leads to a wider strain distribution, characterized by an overall subsidence or extrusion deformation, rather than localized concentration.

(3) Asymmetry in Strain Distribution: When the tunnel intersects a fault, the width of the fault strongly influences the asymmetric distribution of strain in the surrounding rock. If one side of the tunnel is adjacent to the fault, strain concentration becomes more pronounced. In narrow faults, this strain asymmetry is more extreme, with greater differences in strain on either side of the tunnel. In contrast, wide faults tend to result in a more uniform strain distribution around the tunnel, although significant strain differences between the inside and outside of the fault zone persist.

(4) Support Design Considerations: The strain concentration caused by narrow faults may necessitate stronger tunnel support structures near the fault zone. High strain in localized areas may require additional anchors or shotcrete reinforcement to mitigate instability risks in the surrounding rock. Conversely, wide faults produce a broader strain distribution, meaning support design must account for the overall deformation characteristics of the fault zone. Reinforcement measures within wide fault zones, such as reinforced concrete structures or grouting, may be necessary to prevent large-scale plastic deformation and instability within the fault.

In conclusion, fault width has a significant impact on the strain distribution in surrounding rock during tunnel blasting. Narrow faults tend to concentrate strain and localize deformation, while wide faults result in a more extensive strain distribution and significant plastic deformation within the fault zone itself. Therefore, when tunneling through faults of varying widths, it is critical to accurately assess the surrounding rock strain, adjust blasting parameters, and design appropriate support systems to ensure the safety and stability of tunnel construction.

3.2. Damage Characteristics of the Tunnel Surrounding Rock in Fault Fracture Zones

The propagation of stress waves is a highly complex process, characterized by reflection and refraction when the waves encounter interfaces between different media or structural boundaries. These interactions modify both the direction of wave propagation and the distribution of energy, potentially inducing tensile or shear damage in the materials involved. To assess the impact of stress wave propagation at the concrete–surrounding rock interface, a shear stress cloud diagram was generated (Figure 6), illustrating the X-Y cross-section located 1 m ahead of the tunnel face, 3 ms after blasting. The figure reveals several key observations: 3 ms after the blast, the stress wave reaches the concrete–surrounding rock interface, creating a significant stress differential between the two materials. The stress distribution along the concrete lining, resulting from surface blasting, is relatively uniform across the vault in the 0° to 180° range. However, the shear stress distribution is more complex, with internal shear stresses observed within the lining structure. The shear stress values range from −0.51 MPa to 0.50 MPa. When the opposing directional shear stresses exceed a critical threshold, slip deformation may occur at the cross-section. This slip can compromise the integrity of the concrete–surrounding rock interface, potentially reducing the stability of the concrete lining.

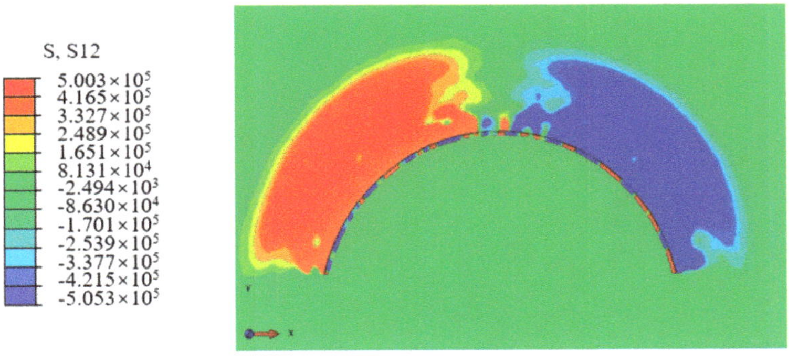

Figure 6. Shear stress cloud of X-Y section.

To assess the impact of stress differentials resulting from variations in body wave velocities across different media, a simulation was conducted. In this simulation, the JH-2 parameters for the surrounding rock material were adjusted to represent the characteristics of weathered granite, while the shear modulus of the lining structure remained unchanged. All other parameters were kept consistent. The resulting stress cloud diagram of the lining structure is shown in Figure 7. The figure indicates that the peak stress decreased from 15.86 MPa to 13.68 MPa, reflecting a reduction of 13.75%. The stress cloud diagram also shows that the C30 concrete support structure experiences rapid stress loading, with stress diffusion occurring at the stress wave interface. In contrast, the structure incorporating the modified rock material exhibits a more distinct stress wave interface, with minimal stress wave diffusion into the surrounding rock mass.

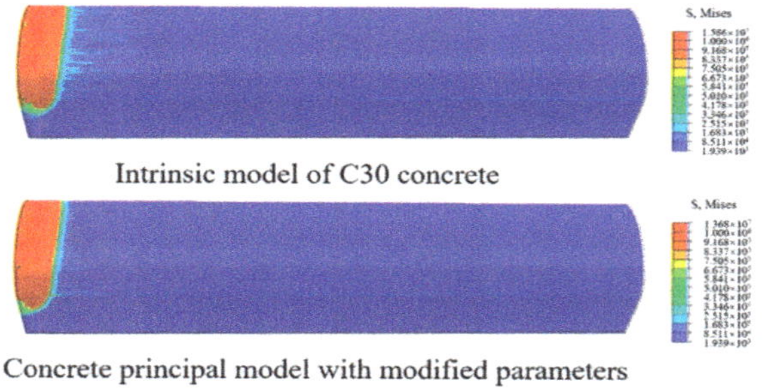

Figure 7. Stress diagram of lining structure.

3.3. Damage Characteristics of the Tunnel Lining Structures in Fault Fracture Zones

Figure 8 presents the circumferential strain curve of the tunnel lining under blasting conditions, with a fault width of 20 m and measured at distances of 1 m, 10 m, and 20 m from the tunnel face. In this diagram, positive strain values indicate deformation toward the interior of the tunnel, while negative values reflect deformation toward the exterior. The following observations are made from the figure:

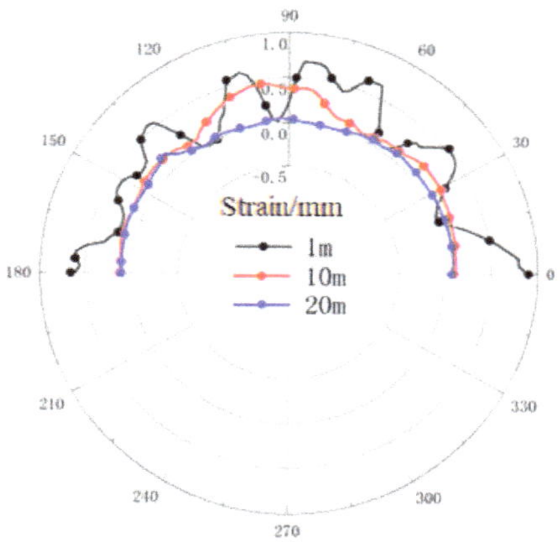

Figure 8. Circumferential strain curves of the tunnel blast lining structure with the distance of 1 m, 10 m, and 20 m from the palm face under the condition of 20 m fault width.

(1) 1 m from the Tunnel Face: At a distance of 1 m from the tunnel face, the lining arch experiences significant circumferential strain, primarily manifesting as compressive strain. This is due to the local stress concentration caused by the blasting-induced stress wave, which particularly affects the tunnel vault. The transient deformation of the surrounding rock results in compressive forces on the lining in the circumferential direction. If the lining design is insufficient to handle this strain concentration, cracks or localized failure may occur at the vault, compromising the structural integrity of the tunnel.

(2) 10 m from the Tunnel Face: At 10 m from the tunnel face, the circumferential strain is reduced compared to that at 1 m due to the attenuation of blast energy as it propagates through the rock mass. However, because of the large fault width and the highly fragmented nature of the surrounding rock, the tunnel lining is still subject to uneven pressure from the surrounding rock. This can result in localized strain concentrations at the arch of the tunnel, with the strain potentially shifting from compressive to tensile. This shift is especially pronounced when the blast stress wave interacts with the fragmented fault zone, which may weaken the overall stability of the lining structure.

(3) 20 m from the Tunnel Face: As the distance from the tunnel face increases to 20 m, the circumferential strain at the vault decreases further, and the deformation begins to stabilize. The long-term impact of the fault width on the lining structure manifests as gradual settlement and plastic deformation. Although the stress levels are lower at this distance and mainly present as uniformly distributed compressive strain or minor tensile strain, the presence of fractured rock within the fault zone still poses a significant risk of deformation, particularly near the fault edges.

In conclusion, the circumferential strain in the tunnel lining is highly influenced by the proximity to the tunnel face and the fault conditions. Close to the tunnel face, compressive strains dominate due to stress concentration, while further away, the strain reduces but becomes uneven due to the fault's fragmentation. The design of the tunnel lining must account for these variations to prevent structural damage, especially near the fault zone.

Characterizing tunnel arch damage is crucial, as the arch often experiences the highest stress concentration, particularly in fault zones or areas with disturbed surrounding rock. The arch serves as a primary load-bearing structure, supporting the overhead load from the surrounding rock. Damage to this component can compromise the tunnel's structural integrity, increasing the risk of progressive collapse or failure. During blasting, the arch is more susceptible to upward and lateral energy waves, which heighten the likelihood of fractures and spalling. In faulted zones, fractured rock may exert uneven pressure on the arch, intensifying stress points and aggravating damage. Such damage can significantly impact the tunnel's overall stability. By focusing on characterizing and mitigating arch damage, engineers can enhance the tunnel's capacity to sustain its designed loads and environmental conditions, ensuring both operational functionality and structural longevity.

Figure 9 illustrates the damage characteristics of the tunnel vault caused by blasting under fault zone conditions with a fault width of 20 m. Several key observations can be made from the figure:

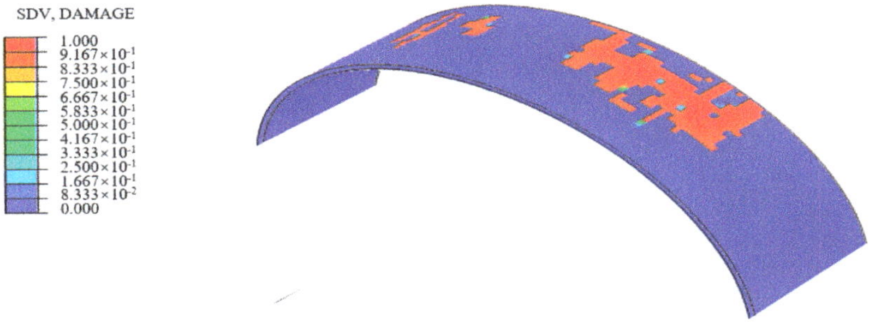

Figure 9. Characteristics of damage and destruction of the vault caused by tunnel blasting under the condition of fault width of 20 m.

(1) Stress Concentration Near the Tunnel Face: In the vicinity of the tunnel face, blasting generates a strong shock wave. This effect is particularly pronounced near fault zones, where the blasting energy leads to stress concentration. As the vault is a critical load-bearing element, it experiences significant circumferential stress, making it highly susceptible to stress concentration during the initial stages of blasting. This can lead to the initiation and propagation of micro-cracks and fractures in the early phases of tunnel excavation.

(2) Impact of Shock Wave Propagation in Fault Zones: As the shock wave travels through the fault zone, the fragmented nature of the rock disrupts the energy propagation path. This can subject the lining structure, particularly the vault, to increased dynamic loads. The high-frequency vibrations generated by the blast contribute to fatigue damage in the vault lining, causing microcracks to gradually expand. Over time, this can lead to crack propagation or material failure if left unaddressed.

In summary, with a fault width of 20 m, the primary damage mechanisms affecting the tunnel vault during blasting include local cracking due to stress concentration, spalling, and fragmentation induced by shock waves. To mitigate these risks, it is essential to enhance the tunnel lining design and implement reinforcement measures, such as anchoring and grouting, to address the complex geological conditions within the fault zone and the impact of blasting shock waves.

4. Conclusions and Recommendations

Tunnel construction in mountainous regions typically employs the blasting method. However, when tunneling through fault fracture zones, this method often results in significant damage to the surrounding rock. To address the geological hazards associated with tunnel blasting in faulted mountain areas, this thesis utilizes a large-scale finite element numerical simulation based on the JH-2 model. The study investigates the damage characteristics of the surrounding rock induced by blasting in fault-fragmented zones, with the Lianfeng Tunnel Project on the Dayong Expressway in Yunnan Province, China, serving as the case study. The findings provide a crucial theoretical foundation for the design and construction of mountain tunnels. The main conclusions of the study are as follows:

(1) Effect of Fault Width on Surrounding Rock Stress: The width of the fault has a significant influence on the stress changes in the surrounding rock during tunnel blasting. Narrow faults tend to cause stress concentrations, increasing the risk of localized rock bursts, while wide faults result in stress wave attenuation, potential fault zone slip, and instability over a broader area of the surrounding rock.

(2) Impact of Fault Width on Surrounding Rock Strain: The fault width also plays a critical role in determining the strain distribution in the surrounding rock. Narrow faults lead to localized strain concentrations and intense deformation, while wide faults cause a more extensive strain distribution, with significant plastic deformation occurring within the fault zone.

(3) Stress waves generated during blasting create stress differentials at the concrete–surrounding rock interface, leading to a complex shear stress distribution and potentially triggering slip-related damage. Utilizing lining materials with strength characteristics similar to those of the surrounding rock can effectively minimize damage at the cemented interface and enhance the overall structural stability of the tunnel.

(4) Vault Damage in Wide Fault Zones (20 m Width): When the fault width is 20 m, the primary damage characteristics of the tunnel vault include local cracking due to stress concentration, spalling, and fragmentation caused by the blasting shock wave, as well as settlement and localized collapse.

To mitigate damage and impact on the surrounding rock and supporting structures in tunnel blasting projects near fault fracture zones, specific countermeasures can be implemented. These strategies focus on enhancing stability, controlling destruction extent, and minimizing impacts on supporting structures. Key measures include controlled blasting techniques, pre-reinforcement of fault zones, enhanced monitoring and early warning systems, dynamic design adjustments, and strengthening of support structures. By integrating these countermeasures, tunnel projects can effectively manage and reduce blasting damage in fault fracture zones, thereby improving safety and extending the lifespan of supporting structures.

Author Contributions: Conceptualization, L.C.; Methodology, F.L.; Software, F.L.; Formal analysis, L.C.; Investigation, P.Z.; Resources, P.Z.; Data curation, F.L.; Writing—original draft, L.C.; Writing—review & editing, Z.Y.; Supervision, Z.Y. and P.Z. All authors have read and agreed to the published version of the manuscript.

Funding: This work was supported by Yunnan Asef-Han Academician Workstation Funded Programs (202105AF150076), Yunnan Provincial Key R&D Program Funded Projects (202003AC10002), and Yunnan Provincial Basic Research Program (202001AT070043).

Data Availability Statement: The original contributions presented in the study are included in the article, further inquiries can be directed to the corresponding author.

Conflicts of Interest: Author Ping Zhao was employed by the company Shandong Province Transportation Planning and Design Institute Group Co. The remaining authors declare that the research was conducted in the absence of any commercial or financial relationships that could be construed as a potential conflict of interest.

References

1. Xiao, K.; Zhou, C.; Zheng, X.; Xu, J. Blasting vibration characteristics and safety criteria of tunnel secondary lining under the influence of fault zone. *Eng. Blasting* **2021**, *27*, 130–139.
2. Lyu, G.; Zhou, C. Damage characteristics of grouted tunnel rock mass in fault zones induced by blasting. *Chin. J. Rock Mech. Eng.* **2021**, *40*, 2038–2047.
3. Wu, X.D.; Miao, X.M.; Gong, M.; Su, J.P.; Zhu, Y.Q.; Chen, X.L. Vibration safety threshold and control technology for blasting to prevent seawater intrusion in coastal tunnel sections near faults. *J. Mar. Sci. Eng.* **2024**, *12*, 1646. [CrossRef]
4. Sha, P.; Wu, F.Q.; Guo, Q.L. Anisotropic deformation in Muzhailing railway tunnel, Gansu, China. In Proceedings of the 3rd International-Society-for-Rock-Mechanics (ISRM) SINOROCK Symposium, Shanghai, China, 18–20 June 2013; pp. 855–860.
5. Yu, Y.Q.; Lu, Y.B.; Lu, Y.P. Numerical simulation of controlled perimeter blasting and field test on the fracture zones in Guanjiao tunnel. In Proceedings of the 2nd ISRM International Young Scholars' Symposium on Rock Mechanics, Beijing, China, 14–16 October 2012; pp. 57–63.
6. Ji, L.; Zhou, C.B.; Lu, S.W.; Jiang, N.; Gutierrez, M. Numerical Studies on the Cumulative Damage Effects and Safety Criterion of a Large Cross-section Tunnel Induced by Single and Multiple Full-Scale Blasting. *Rock Mech. Rock Eng.* **2021**, *54*, 6393–6411. [CrossRef]
7. Wang, M.; Liu, D.; Lu, X.; Hou, X. Blasting Damage of Large Section Tunnel to Surrounding Rock in Near Area. *Trans. Beijing Inst. Technol.* **2024**, *44*, 597–605.
8. Huang, J.; Chen, S.H.; Liu, M.L.; Li, K.P. Physical model test and numerical simulation study of cumulative damage to deep tunnel surrounding rock under cyclic blasting load. *Int. J. Damage Mech.* **2022**, *32*, 161–184. [CrossRef]
9. Liu, X.; Gong, M.; Yang, R.; Wu, X.; Wang, S.; Chen, X. Numerical analysis of surrounding rock damage caused by millisecond delay blasting of tunnel peripheral holes. *J. Vib. Shock* **2023**, *42*, 8–15.
10. Liu, Z.; Wu, J.; Cao, C.; Li, S.T.; Yan, Q.S. Dynamic performance and damage assessment of a shallow buried tunnel under internal explosion. *Tunn. Undergr. Space Technol.* **2023**, *133*, 104918. [CrossRef]
11. Qin, G.; Zeng, C.; Xu, J.; He, R. Numerical Simulation and Engineering Verification of Smooth Blasting in Limestone Tunnel Based on HJC Damage Constitutive Model. *Explos. Mater.* **2023**, *5*, 45–51.
12. Kury, J.W.; Hornig, H.C.; Lee, E.L.; McDonnel, J.L.; Ornellas, D.L.; Finger, M.; Strange, F.M.; Wilkins, M.L. Metal Acceleration by Chemical Explosives. In Proceedings of the 4th Symposium (International) on Detonation, White Oak, MD, USA, 12–15 October 1965; pp. 12–15.
13. Lee, E.; Finger, M.; Collins, W. *JWL Equation of State Coefficients for High Explosives*; Lawrence Livermore National Lab. (LLNL): Livermore, CA, USA, 1973.
14. Li, F.T.; Wu, K.; Li, S.R.; Wang, C.; Liu, Y.J.; Dou, Z.Y. Study on the dynamic response characteristics of lining structures in large-section tunnel blasting using JH-2 model analysis. *Sci. Rep.* **2024**, *14*, 10506. [CrossRef] [PubMed]
15. Liu, J.J.; Li, Y.; Zhang, H.J. Study on Shale's Dynamic Damage Constitutive Model Based on Statistical Distribution. *Shock. Vib.* **2015**, *2015*, 286097. [CrossRef]
16. Yang, Z.; Zhao, Q.; Gan, J.; Zhang, J.; Chen, M.; Zhu, Y. Damage evolution characteristics of siliceous slate with varying initial water content during freeze-thaw cycles. *Sci. Total Environ.* **2024**, *950*, 175200. [CrossRef] [PubMed]
17. Li, X.; Liu, K.; Yang, J.; Song, R. Numerical study on blast-induced fragmentation in deep rock mass. *Int. J. Impact Eng.* **2022**, *170*, 104367. [CrossRef]

buildings

MDPI

Article

Deformation Characteristics of Surrounding Rock of Marine Soft Soil Tunnel Under Cyclic Loading

Wenbin Xu, Yajun Liu, Ke Wu *, Heng Zhang, Yindong Sun and Wenbin Xiao

School of Civil Engineering, Shandong University, Jinan 250061, China
* Correspondence: wuke@sdu.edu.cn

Abstract: Soft marine soil exhibits unique mechanical properties that can lead to significant deformation and instability in the surrounding rock of urban subway tunnels. This presents a critical challenge for tunnel engineering researchers and designers. This thesis investigates the stability characteristics of surrounding rock in marine soft soil tunnels under cyclic loading conditions. Focusing on the shield tunnel segment between Left Fortress Station and Taiziwan Station of Shenzhen Urban Rail Transit Line 12, a discrete–continuous coupled numerical analysis method is employed to examine the deformation characteristics of the surrounding rock. This analysis takes into account the effects of dynamic loads resulting from train operations on the arch bottom's surrounding rock. The findings indicate that damage to the surrounding rock occurs gradually, with the marine soft soil layer, particularly at higher water content, being prone to substantial plastic deformation. Additionally, under the influence of train vibration loads, the degree of vertical fluctuation in the internal marine soft soil diminishes with increasing depth from the bottom of the tunnel arch. This coupled numerical analysis approach offers valuable insights and methodologies for assessing the structural safety of tunnel projects throughout their operational periods.

Keywords: deformation characteristics; discrete–continuous coupled; marine soft soil; numerical analysis; tunnel

1. Introduction

China's coastline extends approximately 18,000 km, and the coastal economy plays a vital role in the nation's overall economic landscape. Coastal cities have been pivotal in driving rapid economic and social development in recent years. The expansion of underground space and the construction of rail transit systems in these urban areas not only enhance transportation efficiency but also serve as key indicators of a country's technological advancement and overall strength in transportation infrastructure. However, engineers and builders encounter unique challenges when developing underground space and rail transit systems in coastal cities. The prevalence of marine soft soils in these regions poses significant obstacles due to their distinct engineering properties, including low shear strength, high compressibility, and high water sensitivity [1,2]. Subway construction on these soft soil foundations is particularly vulnerable to foundation instability and structural damage, raising serious concerns regarding the safety and reliability of such projects. In response to these challenges, the discrete–continuous coupled numerical analysis method offers innovative approaches to understanding the stability characteristics of marine soft soil tunnels under cyclic loading conditions.

Currently, most numerical computational analysis methods, such as the finite element method (FEM) and finite difference method (FDM), are commonly employed to study the stability of marine soft soil tunnels. However, with advancements in computational technology, the discrete element method (DEM), which effectively simulates the behavior of granular materials and complex physical phenomena, has emerged as a powerful tool for analyzing the dynamics of marine soft soils [3]. A numerical method utilizing an elasto-plastic constitutive model that characterizes the mechanical properties and structural features of natural clays has been employed to investigate how additional loads impact soil and tunnel displacements in soft ground layers [4]. Furthermore, the risk of insufficient bearing capacity in soft marine soils during tunnel boring machine (TBM) operations has been examined through analytical equations and three-dimensional numerical analyses that simulate dynamic conditions associated with TBM propulsion [5]. In Nansha, Guangzhou, theoretical and numerical analyses, complemented by monitoring data, have been used to investigate the causes of formation collapse during the construction of hydropower tunnels in marine soft ground strata, leading to the proposal of relevant countermeasures [6]. While the aforementioned studies predominantly utilize continuous element numerical methods, discrete element analysis also plays a critical role in various scientific research fields. For instance, Jian et al. [7] established particle flow code (PFC) models for sandy soils and clays, focusing on different particle contact relationships to conduct preliminary studies on soil mechanical properties. These studies demonstrated the effectiveness of PFC in simulating the mechanical behavior of granular materials and provided valuable insights for both theoretical and practical applications in geotechnical engineering. Additionally, a multi-scale coupling analysis scheme that integrates three-dimensional PFC with fast Lagrangian analysis of continua (PFC-FLAC) has been developed to simulate the earth pressure balance (EPB) shield tunneling process, proposing a method for evaluating face stability based on the velocity state of soil particles entering the soil tank [8]. The safety and stability of existing structures during shallow tunnel excavation in urban areas have also been analyzed using the discrete element method to model the motion and interaction of circular particles in soft marine soils [9]. Moreover, numerical methods employing the elasto-plastic constitutive Shanghai model have been utilized to simulate soil responses and assess the effects of repeated surface surcharge loads on soil and tunnel displacements [10]. Despite the significant advantages of particle flow methods and related software in studying geotechnical engineering properties, there remains a notable gap in their application to the stability analysis of marine soft soil tunnels [11].

While extensive research has been conducted both domestically and internationally on the mechanical properties of marine soft soil and the deformation characteristics of tunnel surrounding rock, there is a relative scarcity of studies that examine the deformation and damage mechanisms of surrounding rock in marine soft soil using the discrete element method in detail. To fill this gap, this study draws on a typical subway tunnel project and employs a discrete–continuous coupled numerical analysis method to investigate the deformation behavior of surrounding rock within marine soft soil layers. The findings offer novel research methodologies and insights into the stability of peripheral rock in marine soft soil tunnels, providing a valuable theoretical foundation for the design and construction of subway tunnels in Chinese coastal cities.

2. Numerical Calculation Models and Analysis Methods

2.1. Project Overview

Shenzhen is located in the Pearl River Delta region, an area characterized by extensive marine-phase soft soils, which pose significant challenges to local engineering projects. The construction of rail transit systems in Shenzhen is particularly affected by these soft soils. This paper focuses on the shield tunnel section between Left Fortress Station and Prince Bay Station on Shenzhen Metro Line 12, investigating the deformation mechanism of marine-phase soft soil in this area. The interval between Left Fortress Station and Prince Bay Station passes under the sea, where the presence of silty clay with low shear strength, high compressibility, and high water sensitivity poses serious risks to project safety. Specifically, the section traverses the marine area near Sanwan Wharf, where the excavation process is significantly influenced by the properties of the soft soil. These conditions increase the likelihood of foundation instability and structural damage, presenting a considerable threat to the safety and reliability of the tunnel construction.

Figure 1 presents a schematic diagram illustrating the distribution of geotechnical strata along the tunnel shield interval from Left Fortress Station to Prince Bay Station, passing through the Sanwan Ferry Terminal area. The primary geotechnical strata in this region consist of vegetal fill, silt, marine soft soil, strongly weathered granite, and slightly weathered granite. Additionally, there is hydraulic connectivity between groundwater and seawater in this area. The shield tunnel crosses the Sanwan Ferry Terminal area, predominantly through the marine-phase soft soil stratum.

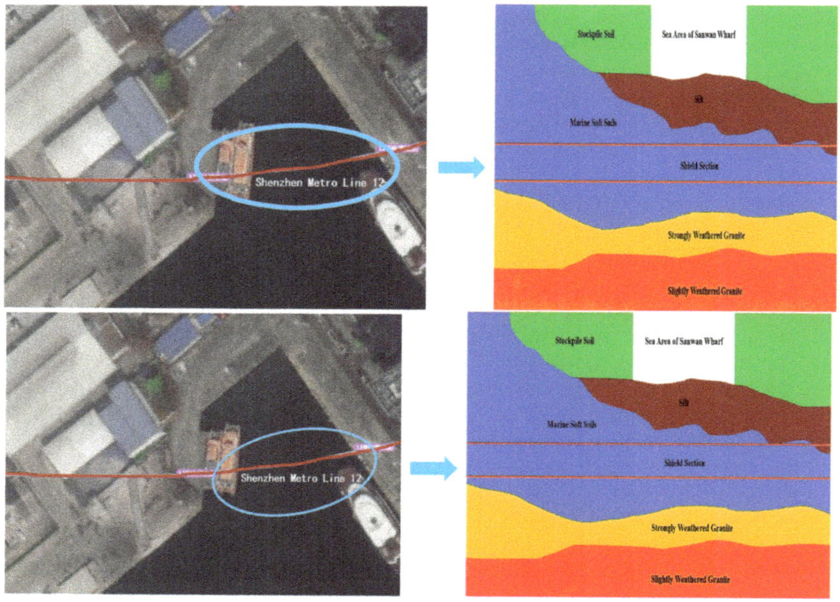

Figure 1. Schematic diagram of the distribution of geotechnical strata.

The left line of the Zuotai interval of Line 12 has a total length of approximately 908.96 m, while the right line measures about 913.18 m. Notably, the left line features approximately 158.56 m of underwater alignment, whereas the right line extends about 185.49 m underwater. The construction of this interval is carried out using the shield method, with two sets of mud–water balanced shield machines tasked with the excavation.

2.2. PFC2D and FLAC2D Coupling Principle

Fast Lagrangian Analysis of Continua (FLAC) is a numerical modeling tool based on the explicit finite difference method, with its primary advantage being the ability to analyze the macroscopic mechanical behavior of continuous media. FLAC is capable of simulating the nonlinear, time-dependent behavior of materials, as well as complex contact and interfacial problems, particularly in soils and rocks. It offers a comprehensive library of material models, supporting a range of behaviors from simple elasticity to complex plasticity and creep. A key feature of FLAC is its high degree of customizability through the FISH programming language, allowing users to develop custom models or analytical functions as needed. In contrast, the Discrete Element Method (DEM) is well suited for solving problems involving discontinuous media. Combining FLAC with DEM leverages the strengths of both methods. For example, in tunnel excavation simulations, the discrete elements can model the soil around the tunnel, while continuous elements represent the intact continuum further from the excavation. This approach allows for detailed observation of soil collapse and failure, while optimizing computational time and maintaining accuracy [12–14].

2.3. Modeling of Discrete–Continuous Coupling

To accurately simulate the engineering site conditions, a two-dimensional discrete–continuous coupling model with dimensions of 80 m × 80 m was established. The tunnel, modeled as a circular structure with a diameter of 4 m, is positioned at a depth of 15 m. Discrete particles are generated within a 20 m × 20 m area surrounding the tunnel, with the surrounding strata represented by PFC2D particle elements. The remaining strata and tunnel structure are modeled using FLAC solid elements.

The boundary conditions of the model are as follows: horizontal displacement is constrained at the left and right boundaries, while vertical displacement at the bottom is fixed. Figure 2 presents the numerical model. The continuous domain is based on the Mohr–Coulomb constitutive model, with stratigraphic parameters drawn from geological data of similar near-coastal engineering projects [3], as outlined in Table 1. For tunnel support, continuous elements simulate concrete lining segments of varying strengths, with elasticity assumed for the lining materials. The strength and parameter values for the support segments are listed in Table 2.

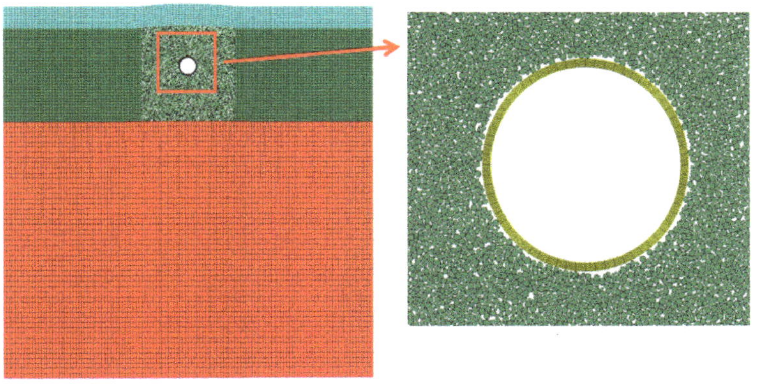

Figure 2. Coupling model.

Table 1. Parameters of each layer.

Stratigraphic Name	Average Thickness (m)	Severe (kN/m³)	Cohesive Force (kPa)	Internal Friction Angle (°)	Elastic Modulus (MPa)
Stockpile Soil	5	18	10	9	12
Marine soft soils	20	16.5	10	5	6
Strongly weathered rocks	55	19.5	18	15	25

Table 2. Strength and parameter values of concrete support pipe segments.

Concrete Type	Severe (kN/m³)	Elastic Modulus (GPa)	Poisson's Ratio
C30	25	31.5	0.2
C40	25	33.5	0.2
C50	25	35.5	0.2
C60	25	37.5	0.2

In the PFC particle flow model, specific parameters for the particle units cannot be directly assigned to replicate the macroscopic properties of the soil. Instead, it is necessary to establish a correlation between the model's fine-scale parameters and the actual physical and mechanical properties of the soil. This requires an iterative process of adjusting the fine-scale parameters until the numerical simulation results accurately reflect the macroscopic physical and mechanical behavior of the soil—this process is known as macro–fine parameter calibration. The key microscopic parameters in the linear contact bond model include the friction coefficient, linear contact effective modulus, the ratio of normal to tangential contact stiffness between particles, normal bond strength, and tangential bond strength. By fine-tuning these microscopic parameters, the numerical results can be brought into closer agreement with the results from laboratory tests. The calibrated fine-scale parameters are listed in Table 3 [3].

Table 3. Calibration results of microscopic parameters.

Sample State	Coefficient of Friction	Linear Contact Effective Modulus (MPa)	Rigidity Ratio	Normal Bond Strength (kPa)	Tangential Bond Strength (kPa)
Water content 40%	0.17	50	2	30	30
Water content 45%	0.15	40	2	25	25
Water content 50%	0.14	20	2	20	20
Water content 55%	0.13	10	2	10	10
Water content 60%	0.11	8	2	1	1

During the research process of this paper, the influence of various parameters in Tables 1–3 on numerical simulation has been discussed in related papers [15]. In this paper, we focus on the deformation characteristics of the surrounding rock of the sea-phase soft soil tunnel under the action of train dynamic load, in which the dynamic load and water content are the two parameters focused on the discussion, and other parameter variability is not considered for the time being.

In this study, a scaled-down model based on the centrifuge principle was used, with a reduction ratio of 1:100. This scaling significantly reduces the number of particles in the model, bringing the total number of particles in the discrete domain down to 42,676, which effectively controls the computational complexity while maintaining a high level of detail. The use of this reduced-scale model not only enhances computational efficiency but also enables accurate simulation of the deformation characteristics of the ground surrounding the tunnel, even with limited computational resources.

To ensure that the mechanical response of the scaled model corresponds to that of the full-scale model, the stress levels within the strata are kept constant. Given the reduced dimensions of the scaled model, the gravitational acceleration is increased by a factor of 100 to match the stress conditions in the actual setting. This approach has been widely used in similar simulations [16]. Although the model is scaled down, the physical and mechanical properties of the materials, such as elastic modulus and Poisson's ratio, remain unchanged. This adjustment ensures that the stress state in the formation is consistent, allowing the scaled model to accurately represent the behavior and response of the soil under real-world stress conditions.

3. Results

3.1. Deformation Characteristics of Particle Flow in Tunnel Surrounding Rock

The failure of the surrounding rock is a gradual process, and the destabilized zones of the surrounding rock at various time steps under unsupported conditions are shown in Figure 3. From the figures, it is evident that the coupled model effectively simulates the tunnel excavation process without support. At the 1000th time step, the destabilized zone is concentrated around the tunnel arch. In the early stages of excavation, the arch particles experience contact failure due to the influence of self-weight. Under unsupported conditions, a temporary pressure arch forms in the surrounding rock, providing brief stability. As time progresses and the stress field evolves, the destabilized zone begins to expand outward. Its shape gradually develops into an ellipsoid with the vertical axis as the long axis. By the 6000th time step, the destabilized zone has extended to the surface, indicating that the range and severity of destabilization increase with continued excavation.

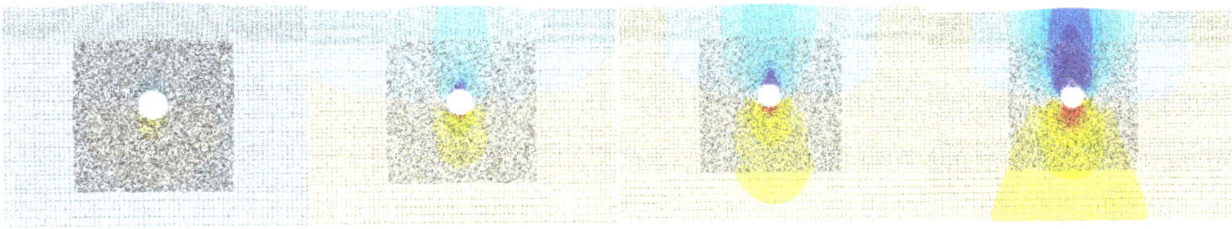

(a) cycle 1000 (b) cycle 2000 (c) cycle 3000 (d) cycle 6000

Figure 3. Distribution of surrounding rock instability zone in unsupported state of coupled model.

Liang Xiaodan et al. [17] defined the location where the surrounding rock's shear stress returns to the original rock stress level as the inner boundary of the pressure arch. Under specific conditions, after tunnel excavation, the surrounding rock can achieve a stable equilibrium state through stress redistribution. In the case of unsupported excavation, the variation of shear stress in the surrounding rock is measured through the simulation process, and the resulting shear stress change curves at different distances from the top of the arch are shown in Figure 4. As depicted in the figure, the shear stress in the surrounding rock within 2 m of the arch decreases rapidly over a short period, indicating that this section of rock is bearing significant loads during the initial stages of tunnel excavation. This emphasizes the importance of ensuring that support measures applied during excavation can withstand these initial loads. As the simulation progresses, the shear stress in the surrounding rock at 10 m above the arch and beyond gradually decreases, eventually extending to the surface.

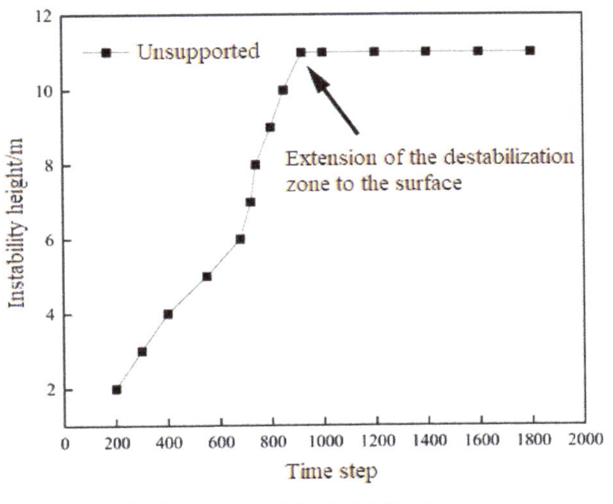

Figure 4. Height change curve of the destabilization zone.

Figure 5 illustrates the force chain diagram of particles after excavation. In the PFC2D-generated force chain diagram, black line segments represent contact forces in compression, with the width of each segment indicating the magnitude of the contact force. As shown in the figure, the force chains in the vault area are relatively sparse, indicating lower contact forces in this region. In contrast, the force chains in the arch waist area are wider and denser, reflecting higher contact forces. Analyzing the force transfer paths reveals that, prior to excavation, the force was primarily transferred vertically downward. After excavation, the force transfer paths significantly deflected around the excavated zone, suggesting that this deflection is the key factor in the formation of the pressure arch. This change in force transfer indicates that the stress state of the surrounding rock was redistributed during excavation, leading to the formation of a new stabilizing structure—the pressure arch—that effectively supports the weight of the overlying soil and other loads.

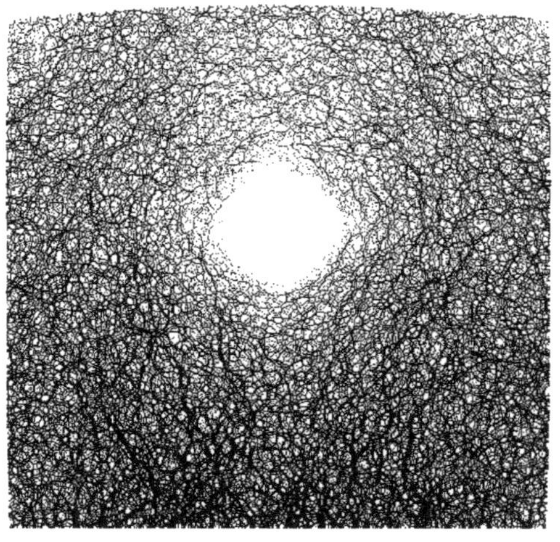

Figure 5. Force chain diagram for discrete particles after tunnel excavation.

3.2. Deformation Characteristics of Tunnel Surrounding Rock Under Different Water Content Conditions

To analyze settlement deformation at a depth of 4 m beneath the tunnel arch in the coupled PFC2D and FLAC2D model, the cumulative deformation of the soil is transformed into positive values and extracted. The cumulative deformation curves under different water content conditions are shown in Figure 6a. These curves are compared with indoor test results under similar water content conditions (Figure 6b) and with discrete element simulation results (Figure 6c). The comparison reveals a clear consistency between the cumulative deformation trends from the coupled model and the results from both the indoor tests and discrete element simulations. This comparison not only confirms the reliability of the coupled model in simulating cumulative deformation at the arch bottom, but also emphasizes the applicability of the results from indoor tests and discrete element simulations within the coupled model. In all approaches—indoor tests, discrete element simulations, and the coupled model—the soil samples exhibited irreversible deformation accumulation under repeated loading conditions. The accumulation of plastic strain was proportional to the water content, with the most significant increases occurring in the early stages of loading. As loading cycles increased, deformation stabilized across all models. These findings suggest that the deformation response at the tunnel arch bottom in the coupled model reflects not only a macroscopic mechanical response, but also the interaction between microstructural changes and the macroscopic mechanical properties of the soil. Indoor tests provide an intuitive understanding of the macroscopic behavior of the soil, while discrete element simulations offer insights into particle-level interactions. The tunnel coupling model integrates these approaches, further validating the deformation characteristics of the soil.

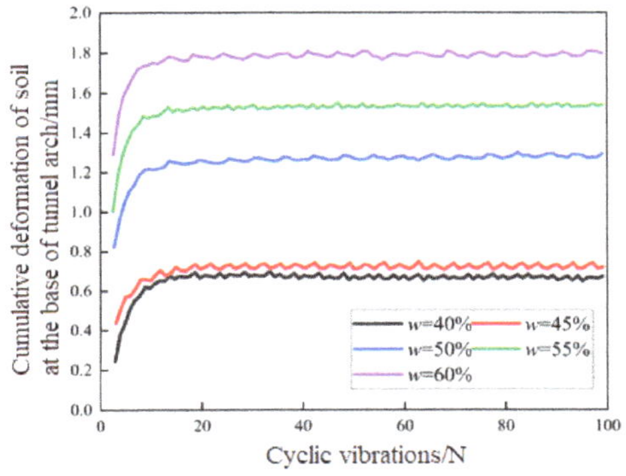

(**a**) Cumulative soil deformation at 4 m at the base of the arch

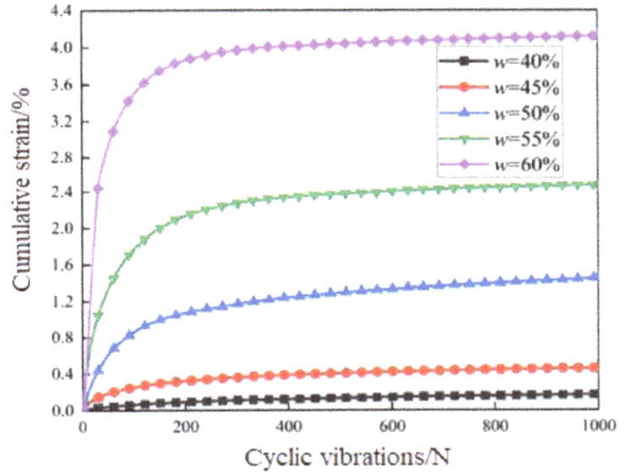

(**b**) Cumulative strain curve of indoor test

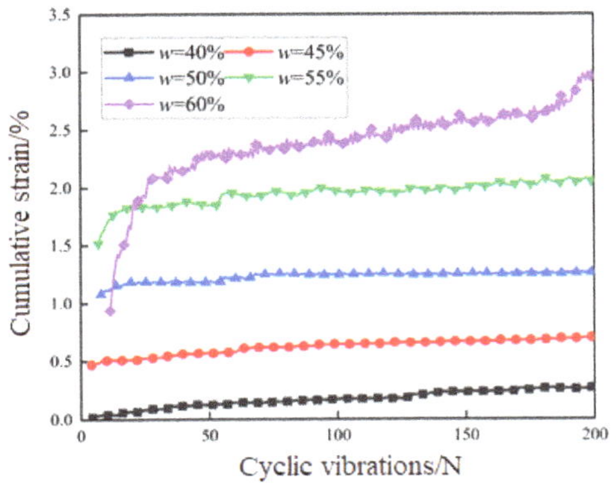

(**c**) Discrete element simulation of cumulative strain curves

Figure 6. Comparison of cumulative deformation of different models.

3.3. Deformation Characteristics of Surrounding Rock at the Bottom of Tunnel Arch Under Cyclic Load

In investigating the stability of marine soft soil tunnels under dynamic train loading, two key factors—water content of the marine soft soil and the distance from the train's point of action to the bottom of the arch—are critical for understanding tunnel stability under such conditions. To explore these factors, relevant measurement points were established within the numerical model, and boundary conditions corresponding to the train load were applied. This allowed for a detailed numerical simulation of the marine soft soil characteristics. The subway vibration load was simulated by applying a velocity at the bottom of the tunnel, with common displacements observed to range from 1 to 3 mm. For the cyclic loading scenario, the vibration displacement at the bottom of the arch was controlled at 2 mm, with a real-time cycle length of 0.04 s, repeated for a total of 100 cycles [18,19]. Additionally, measuring points were established in the soil at the bottom of the arch to analyze the deformation patterns around the tunnel under varying moisture content conditions. The analysis focused on the deformation behavior of the soil surrounding the tunnel under subway vibration loading, as illustrated in Figure 7.

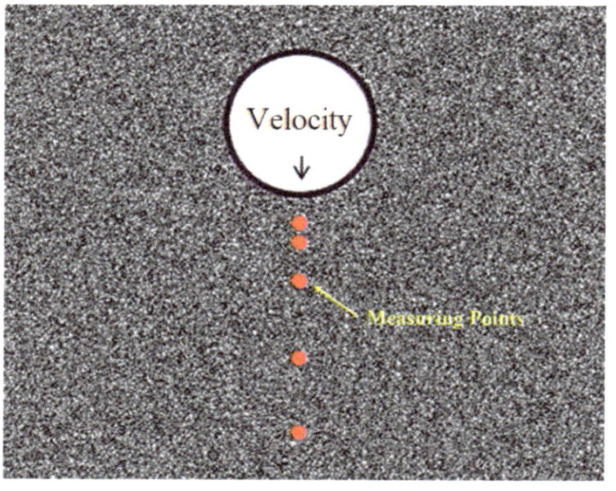

Figure 7. Schematic diagram of measuring point.

Figure 8 shows the displacement curves at different depths of the arch base under 100 cycles of loading conditions. The velocity applied in the simulation is intended to simulate the vibration generated during the operation of the subway, which is able to propagate through the soil body, leading to interactions and displacements between particles. It can be seen that as the number of vibrations increases, the particles within the soil rearranges and produces plastic deformation. The initial vibration may lead to relative displacements between the particles, and as the vibration continues, these displacements accumulate leading to changes in the structure of the soil body and the formation of plastic deformations, and the results obtained are consistent with the results of the triaxial test simulations [3].

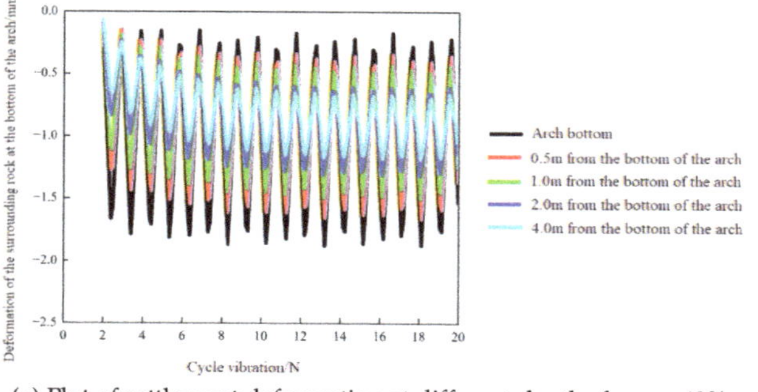

(**a**) Plot of settlement deformation at different depths for w = 40%

Figure 8. *Cont.*

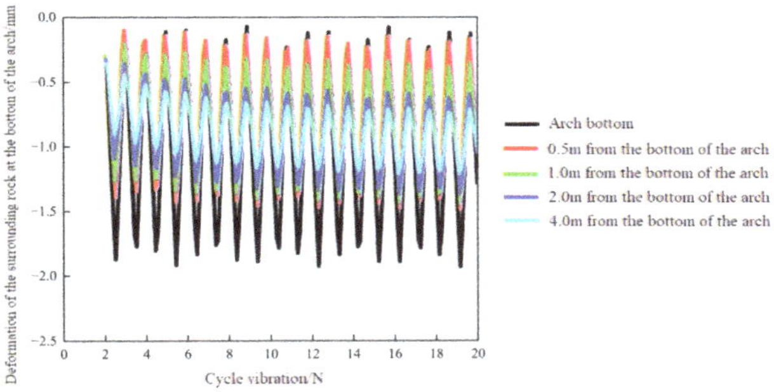

(**b**) Plot of settlement deformation at different depths for w = 45%

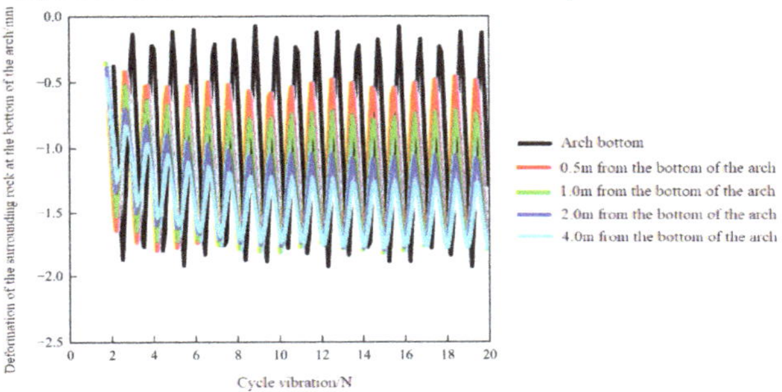

(**c**) Plot of settlement deformation at different depths for w = 50%

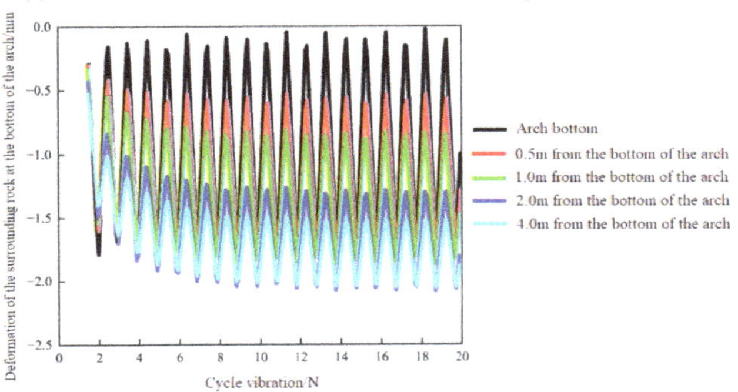

(**d**) Plot of settlement deformation at different depths for w = 55%

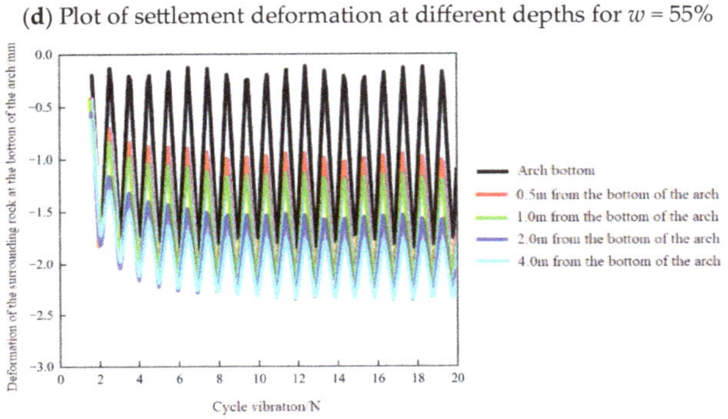

(**e**) Plot of settlement deformation at different depths for w = 60%

Figure 8. Settlement deformation of arch bottom under different moisture content conditions.

The deformation of marine soft soil at the base of the tunnel arch decreases with increasing distance from the arch base, while it increases with higher water content. During cyclic loading, the degree of upward and downward fluctuations in the soft soil diminishes as the depth from the arch base increases. This behavior is attributed to the attenuation of vibration load propagation with depth; particles located deeper within the marine soft soil are less affected by vibrations, resulting in comparatively smaller displacements and deformations. Consequently, the impact of vibration perturbations on the marine soft soil decreases with depth. As the number of vibration cycles increases, the plastic deformation of the soil also rises. This cumulative effect arises because each vibration induces minor displacements in soil particles, which accumulate over multiple cycles, leading to significant plastic deformation. This effect is particularly pronounced under conditions of higher water content, where the reduction in friction between particles further enhances soil susceptibility to deformation under external loads. As water content increases, the shear strength of the soil decreases, making it more prone to larger plastic deformations under the same vibration load.

4. Conclusions

Constructing subway tunnels on marine soft ground frequently poses challenges due to severe foundation deformation. To address this, we developed a discrete–continuous coupled model to investigate tunnel envelope deformation under cyclic train loading, with a particular focus on water content variation. This approach offers an innovative framework for researching and understanding the stability of marine soft soil tunnels. Our findings indicate that the discrete–continuous coupled numerical method is an effective tool for assessing marine soft soil tunnel stability at the microparticle level. This method allows for a more comprehensive and precise understanding of deformation laws in marine soft soil from both macroscopic and microscopic perspectives, providing a valuable theoretical foundation for subway tunnel research and design. Furthermore, our analysis shows that water content and cyclic train loading are critical factors that significantly impact deformation of the tunnel perimeter rock. To mitigate deformation during subway construction, reinforcement measures—such as pre-injection grouting around the tunnel and an overrunning support system within the tunnel—can be implemented to enhance the stability and safety of tunnels under dynamic train loading.

Author Contributions: Formal analysis & investigation, Y.L.; Investigation & data curation, W.X. (Wenbin Xu); Data curation & writing—original draft, H.Z. and Y.S.; Formal analysis & writing—original draft, W.X. (Wenbin Xiao); Writing—review & editing, K.W. All authors have read and agreed to the published version of the manuscript.

Funding: This research was funded by National Natural Science Foundation of China (52179106).

Data Availability Statement: The data presented in this study are available on request from the corresponding author.

Conflicts of Interest: The authors declare no conflict of interest.

References

1. Kongar-Syuryun, C.; Klyuev, R.; Golik, V.; Oganesyan, A.; Solovykh, D.; Khayrutdinov, M.; Adigamov, D. Principles of sustainable development of georesources as a way to reduce urban vulnerability. *Urban Sci.* **2024**, *8*, 44. [CrossRef]
2. Wu, K.; Xu, W.; Liu, Y.; Liu, D.; Dou, Z.; Liu, Z.; Chen, R.; Hao, D. Mechanical properties of marine soft soil and its influence on urban underground track transport engineering. *J. Water Resour. Archit. Eng.* **2024**, *22*, 122–131.
3. Xiao, W.; Wu, K.; Xu, W.; Liu, Y.; Lu, H.; Chen, R. Experiment and analysis on dynamic characteristics of marine soft clay. *Mar. Georesour. Geotechnol.* **2024**, *5*, 1–22. [CrossRef]
4. Xie, W.B.; Zhang, Q.; Zhu, W.X. Numerical investigation on the effect of repeated surface surcharge loading on soil deformation and tunnel displacement in structured soft clay. *Soil Dyn. Earthq. Eng.* **2024**, *181*, 108657. [CrossRef]
5. You, K. A study on the risk and settlement evaluation of a shield TBM excavated in soft marine sedimentary soils. *J. Korean Tunn. Undergr. Space Assoc.* **2016**, *18*, 355–364. [CrossRef]

6. Song, X. Analysis of the deformation and failure mechanism of a utility tunnel excavation in the deep and soft clay area of Nansha, Guangzhou. *Chin. J. Rock Mech. Eng.* **2023**, *42*, 3629–3642.
7. Jian, Z.; Yan, S.; Yong, C. Simulation of soil properties by particle flow code. *Chin. J. Geotech. Eng.* **2006**, *28*, 390–396.
8. Fu, J.; Xie, J.; Fang, Y.; Yang, J. PFC-FLAC coupling analysis of face stability for EPB shield tunneling. *J. Huazhong Univ. Sci. Technol. (Nat. Sci. Ed.)* **2019**, *47*, 116–121.
9. Castelli, F.; Motta, E. Numerical analysis for provision of tunneling-Induced ground deformation in granular soil. In Proceedings of the 5th International Conference of TC28 of the ISSMGE, Amsterdam, The Netherlands, 15–17 June 2005; pp. 503–508.
10. Tian, Y.; Wang, L.; Jin, H.; Xia, Y.C.; Zeng, B. Failure mechanism of horizontal layered rock slope under the coupling of earthquake and weathering. *Adv. Civ. Eng.* **2020**, *2020*, 8825440. [CrossRef]
11. Kongar-Syuryun, C.B.; Kovalski, E.R. Hardening backfill at potash mines: Promising materials regulating stress-strain behavior of rock mass. *Geol. Geophys. Russ. South* **2023**, *13*, 177–187.
12. He, Z.; Xiang, D.; Liu, Y. Triaxial creep test and particle flow simulation of coarse-grained soil embankment filler. *Front. Earth Sci.* **2020**, *8*, 62. [CrossRef]
13. Shao, L.; Chi, S.; Jia, Y. Meso-mechanical simulation of a large scale triaxial test of rockfill materials. *Rock Soil Mech.* **2009**, *30*, 239–243.
14. Powrie, W.; Ni, Q.; Harkness, R.M.; Zhang, X. Numerical modelling of plane strain tests on sands using a particulate approach. *Géotechnique* **2005**, *55*, 297–306. [CrossRef]
15. Xiao, W. Research on Strength and Deformation Characteristics of Marine Soft Soil Around the Tunnel Under Cyclic Load of Subway. Master's Thesis, Shandong University, Jinan, China, 2024.
16. Hou, J.; Zhang, M.; Li, P. Simulation study of foundations reinforced with horizontal-vertical inclusions using particle flow code. *J. Shanghai Jiaotong Univ. (Sci.)* **2013**, *18*, 311–316. [CrossRef]
17. Liang, X.; Zhao, J.; Song, H. Expermimental and numerical analysis of the arching action from stress adjusting in surrounding rocks. *J. Eng. Geol.* **2012**, *20*, 96–102.
18. Wei, Z. Dynamic Characteristics and Long-Term Settlement of Sft Soil Layer Under Vibration of Subway Train. Master's Thesis, Southeast University, Nanjing, China, 2022.
19. Hang, Y. Study on the Deformation Law of Soft Clay Under Dynamic Load of Urban Metro. Master's Thesis, Shanghai University of Applied Sciences, Shanghai, China, 2021.

buildings

MDPI

Article

Study of Semi-Adiabatic Temperature Rise Test of Mineral Admixture Concrete

Ke Wu [1], Zhenhua Liu [1,*], Cao Wang [2], Tao Yang [2], Zhongyu Dou [1] and Jiaxiang Xu [1]

[1] School of Civil Engineering, Shandong University, Jinan 250061, China; wuke@mail.sdu.edu.cn (K.W.); 202335075@mail.sdu.edu.cn (Z.D.)
[2] Power China Municipal Construction Group Co., Ltd., Tianjin 300384, China
* Correspondence: 202315060@mail.sdu.edu.cn

Abstract: The concrete used in the main structures of subway stations has a high degree of constraint. Consequently, temperature changes and shrinkage during construction frequently lead to significant constraint stress, which can result in structural cracking. Therefore, cement with low hydration heat is commonly used in engineering to reduce the temperature of concrete during its age. Aiming at the problem of hydration and heat release caused by concrete construction, based on the principles of concrete hydration heat release and a numerical analysis method, an optimized semi-adiabatic temperature rise test method has been introduced to investigate concrete temperature rise characteristics with different mineral admixtures. The following conclusions were obtained: The effect of reducing the heat of hydration is related to the content and material properties of different mineral admixtures, but not the type of mineral admixture to be incorporated. The temperature rise performance of four common mineral admixtures is as follows: ① total cooling capacity: limestone powder > slag, fly ash > metakaolin; ② early heat generation rate: metakaolin > slag > fly ash > limestone powder; ③ heat reduction rate in the middle and late periods: metakaolin > limestone powder > fly ash > slag.

Keywords: concrete; mineral admixture; temperature rise test; numerical simulation

1. Introduction

With China's economy rapidly advancing, the construction of urban rail transit is developing at a high speed, and subway stations are also developing in larger, more efficient, and international directions [1]. As a vital component of urban rail construction process, the environmental pollution and safety hazards of subway stations that occur in the construction process require great attention. The concrete used in the main structures of subway stations has a high degree of constraint; thus, the temperature effect [2] and shrinkage effect [3] during construction frequently lead to significant constraint stress, which can result in structural cracking. It is commonly believed that early confining stress in concrete mainly arises from two sources: temperature stress due to temperature variations and shrinkage stress resulting from the concrete's contraction [4]. The temperature confining stress is composed of the following two parts: one is the self-constraint stress which arises from the temperature difference between the inside and the surface of the structure, and the other is the external constraint stress caused by the constraints of the adjacent members and the base during cooling and shrinkage. In addition, most of the main structures of subway stations are located underground, and cracks in the station structures during the age period will cause water seepage in the structure because groundwater generally contains chloride ions. Such water seepage problems not only affect the durability and use of the station structure, but also further expand the structural cracking and reduce the structural strength [5]. Therefore, in view of the temperature cracking problem of the primary structure concrete of subway stations, it is particularly important to carry out relevant research to prevent seepage cracks in the structure of subway stations. According to relevant statistics [6–8], the temperature effect, shrinkage effect, and uneven settlement of concrete in the primary structure of subway station are the main causes of cracks in the concrete structure, and 80% of the cracks in the engineering structure are caused by this. Unlike the external load action mode with clear force transmission and clear analysis, the shrinkage and temperature action of concrete structures during the lifetime are uncertain and have various relationships with materials, structural forms, curing conditions, environmental factors, etc. In addition, various thermodynamic and mechanical properties of concrete constantly change with age [9].

As for the relevant theories of concrete cracking, the research methods are mainly divided into those based on fracture mechanics and those based on damage mechanics, which represent macro- and micro-perspectives, respectively. For fracture mechanics, Beyazit B et al. [10] used a non-local fracture simulation tool and adopted the overlapping lattice method to optimize the fracture model, enabling the method to accurately estimate the fracture propagation direction and width. Pasquale Gallo et al. [11] proposed a new brittle fracture formula based on extended strain energy density from macroscopic scale to atomic scale. The results showed that the cracking behavior of ideal brittle substances is independent of the research scale, and it was concluded that the fracture nature of materials is controlled by the fracture of atomic bonds. China's concrete fracture mechanics started in the late 1970s, and the first academic conference on Rock Concrete Fracture and Strength was held in 1981, which promoted the development and research of concrete in the area of fracture mechanics [12]. Zhang Xiufang et al. [13] conducted relevant tests to study the connection between crack length and fracture energy characteristics of concrete. Xu Shi-shin et al. [14] proposed a new wedge-type compact tensile fracture test for reducing the size effect of the phenomenon that additional bending moments are easy to produce in the conventional wedge-splitting test of concrete.

For damage mechanics, Hillerborg et al. [15] in Sweden established a virtual crack model that can reflect the softening characteristics of concrete after crack failure. Bazant et al. [16] successively proposed fracture models with similar concepts as "virtual fracture", such as "fracture zone", "separation fracture", and "distributed fracture", all of which can be used for finite element calculation. Jun et al. [17] developed a constitutive model for concrete using damage theory from the viewpoint of composite materials and analyzed its damage and fracture behavior through numerical simulations. Wang Licheng et al. [18] regarded concrete as a three-phase complex composed of mortar, aggregate, and the interfacial transition zone (ITZ) between them, and theoretically analyzed its heat conduction characteristics from a microscopic perspective and made corresponding predictions. Jia Fujie [19] established a finite element analysis calculation model of semi-adiabatic temperature rise test for concrete temperature field calculation, and verified and modified the model in engineering, which provided a reliable basis for the semi-adiabatic temperature rise test of concrete. Yun H. D. et al. [20] studied the prediction method of heat of hydration of fly ash concrete. The authors proposed a new prediction model by combining adiabatic temperature rise test with regression analysis. The results show that the model can accurately predict the hydration heat release process of fly ash concrete under different mix ratios. Liang T. et al. [21] discussed the effect of hydration temperature rise inhibitors on concrete temperature rise and its mechanism. Through experimental research, the authors found that the addition of hydration temperature rise inhibitors can effectively reduce the hydration temperature rise of concrete and significantly slow down the temperature rise rate. Quan J. et al. [22] studied the adiabatic temperature rise characteristics of fly ash concrete with medium content and its modeling method. The authors experimentally measured the adiabatic temperature rise rate of concrete under different fly ash contents, and proposed a temperature rise rate prediction model based on experimental data.

However, in large-scale subway station projects, the difference in temperature between the interior and exterior of the concrete has become more pronounced, and the temperature effect of concrete is often the key factor affecting concrete cracking. In such projects, cement with low hydration heat is generally used to reduce the heat of hydration released by concrete during its age. Therefore, to solve the problem of concrete cracking caused by temperature effect in subway stations, this paper carries out the simulation of a semi-adiabatic temperature rise test based on the principle of concrete hydration and heat release and a numerical analysis method, and proposes an optimized semi-adiabatic temperature rise test method. Compared with the adiabatic temperature rise test method, this method is more in line with engineering practice, which is helpful to explore the temperature rise characteristics of concrete with different mineral admixtures and provides a reliable basis for the selection and mix ratio of concrete admixtures in practical engineering.

2. Semi-Adiabatic Temperature Rise Test and Finite Element Analysis Method of Concrete

2.1. Semi-Adiabatic Temperature Rise Test of Concrete

The adiabatic temperature rise test can accurately measure the adiabatic temperature rise characteristics of concrete, usually using an adiabatic temperature rise test chamber. However, because of relatively expensive adiabatic temperature rise test equipment, many institutions and construction units do not have conditions for testing, so a more convenient and inexpensive semi-adiabatic temperature rise test is increasingly being used. Jia Fujie [19] also verified the accuracy of semi-adiabatic temperature rise tests, indicating that it can reasonably predict the temperature rise characteristics of concrete. In simple terms, the semi-adiabatic temperature rise test monitors the temperature change of the concrete test block by creating a stable temperature change environment for the concrete in the mold, to deduce the adiabatic temperature rise characteristics of the concrete more accurately. The semi-adiabatic temperature rise test device generally consists of two parts: a concrete mold and an incubator. The insulation material is generally polyurethane, and the concrete mold is generally a cube mold. The incubator was composed of insulation material around a concrete mold, as shown in Figure 1.

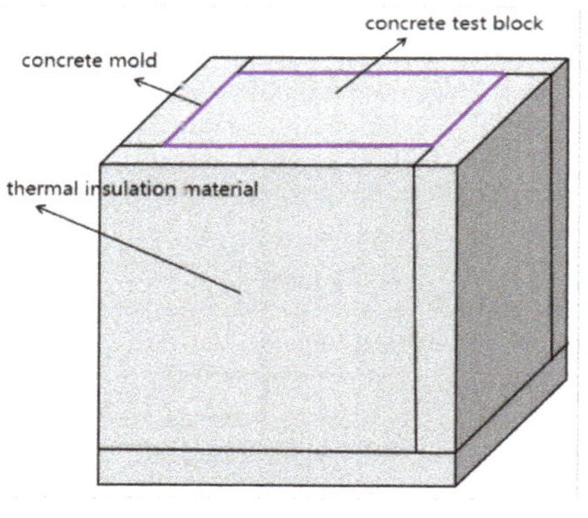

Figure 1. Semi-adiabatic temperature rise test device.

2.2. Finite Element Analysis Method for Hydration Heat Release of Concrete

To optimize the semi-adiabatic temperature rise test and solve the adiabatic temperature rise model of concrete by the method of numerical simulation, it is essential to perform a finite element analysis of the concrete's heat release process scientifically and reasonably. Therefore, it is necessary to master the principle of hydration and heat release of concrete clearly, choose an appropriate hydration and heat release model, and realize it in ABAQUS for finite element analysis. In this study, the HETVAL subroutine of ABAQUS was used for thermogenic analysis. The heat release temperature field analysis of concrete hydration is comparable to placing a heat source inside the concrete, where the heat output changes as it ages. Since ABAQUS lacks a direct function for simulating a time-varying heat source, the HETVAL subroutine must be used to implement this functionality and integrate it into the calculation of the concrete temperature field. Due to the complexity of the hydration heat release curve, engineering often relies on simplified empirical fitting expressions. In this research, the composite exponential model selected earlier is applied to determine the temperature field. To realize concrete hydration heat release in the HETVAL subroutine, the most important thing is to establish the heat release rate, represented by FLUX(1), for the heat source in the subroutine. This rate is determined from Equation (1), and the specific formula is as follows:

$$\text{FLUX}(1) = q_v = \frac{abQ_0}{24}\left(\frac{t}{24}\right)^{b-1} e^{-a(\frac{t}{24})^b} \tag{1}$$

where q_v is the heat release rate of the heat source, Q_0 is the greatest level of heat release from concrete hydration, and a and b are the parameters of the exothermic model.

In the subroutine, the Fortran language is used to write the aforementioned formula, resulting in the assignment of the value q_v to FLUX(1) and storing it in the state variable denoted as statev, facilitating its accessibility for post-processing purposes, such as output generation and viewing. In the heat transfer analysis model, the HETVAL subroutine acts as the heat source. This heat source changes with time, so the time t in q_v is defined as the age of the concrete, and time (2) represents the total time at the end of the increment. According to the finite element analysis method of concrete hydration and heat release, the temperature field of different concrete structures in the age can be determined.

2.3. Finite Element Analysis Model of Semi-Adiabatic Temperature Rise Test Device

The model was built in ABAQUS according to the test device shown in Figure 1. A three-dimensional model of the test device is shown in Figure 2. The model consists of a concrete test block with polyurethane insulation that encloses it. The concrete test block is in the shape of a cube with a side length of 20 mm, and the concrete test block is completely wrapped by an incubator composed of insulation layer.

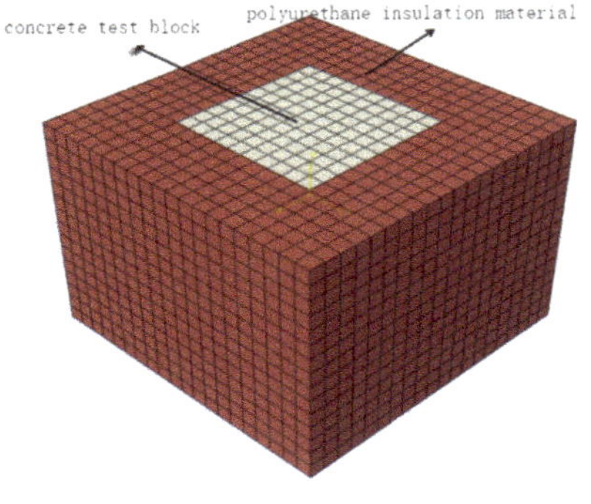

Figure 2. Three-dimensional model of temperature rise test. Reprint with permission [23]; 2024, Applied Sciences.

In ABAQUS, there are three modes of heat transfer in this experiment: (1) Heat conduction, which takes place within the concrete test block, inside the insulation material, and at the interface between the concrete and insulation, is modeled by specifying the thermal conductivity in the attribute module. (2) Heat convection happens at the surface of the insulation layer, where it contacts the external air, managed by specifying the heat dissipation coefficient in the contact module. (3) Thermal radiation takes place during the heat exchange between the entire semi-adiabatic temperature rise device and the surrounding environment. This is controlled by setting the emissivity parameter in the contact module, along with defining the values for absolute zero and the Stefan–Boltzmann constant. The Stefan–Boltzmann constant is used in the Stefan–Boltzmann law, which states that the total amount of energy radiated per unit area of a blackbody surface (called the radiance or energy flux density of the object) is proportional to the fourth power of the thermodynamic temperature of the blackbody itself, where the proportionality coefficient is this constant.

To optimize the test via numerical simulation, the heat release model parameters of ordinary Portland cement No. 525 with high hydrothermal generation were selected according to Table 1 (to better compare the differences in curves), and various properties of the polyurethane insulation material were initially selected. The insulation materials commonly used in projects are polystyrene foam board, extruded plastic board, polyurethane foam board, etc., and the thickness is generally between 5 and 10 cm. This is consistent with the parameter settings for the finite element simulation in this article. The calculation parameters for the test model are listed in Table 1.

Table 1. Calculation parameters of semi-adiabatic temperature rise test model.

Argument	Value
Thermal conductivity of concrete	1.78
Coefficient of thermal expansion of concrete	1.0×10^{-5}
Specific heat of concrete (J/(kg·°C))	960
Concrete mass density (kg/m^2)	2400
Heat release model parameter Q_0 (kJ/kg)	3500
Influence coefficient of hydration heat release model a	0.36
Influence coefficient of hydration heat release model b	0.74
Thermal conductivity of thermal insulation material (W/m·K)	0.024
Specific heat of insulation material (J/(kg·°C))	1200
Insulation material density (kg/m^2)	36
Film heat dissipation coefficient	17,000
Emissivity	0.77
Absolute zero degree Celsius (°C)	−273.15
Stefan–Boltzmann constant	5.669×10^{-14}
Mold entry temperature (°C)	25.5

Based on the above calculation parameters, the semi-adiabatic temperature increase test model was numerically simulated, and a temperature cloud image of the semi-adiabatic temperature increase test device was obtained, as shown in Figure 3. The figure shows that the temperature of the concrete test block under the action of the heat source is much higher than that of the insulation material, and the heat does not penetrate the insulation layer. This is because the heat inside the concrete after being wrapped by thermal insulation material can only be dissipated through the heat conduction between the insulation material and the thermal insulation material. At the same time, the heat absorption and insulation ability of the insulation material is strong, resulting in the temperature of the concrete test block being generally higher than that of the insulation material. The insulation material can provide a stable temperature change environment for the concrete test block.

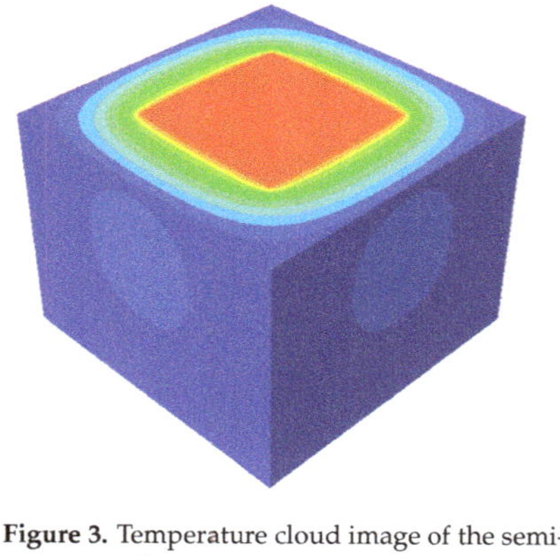

Figure 3. Temperature cloud image of the semi-adiabatic temperature rise test device. Reprint with permission [23]; 2024, Applied Sciences.

3. Optimal Design of Semi-Adiabatic Temperature Rise Test Device

3.1. Influence of Insulation Layer Thickness on Semi-Adiabatic Temperature Rise Test

In the traditional semi-adiabatic temperature rise test, there is no clear regulation on the insulation layer thickness, so the use of thicker insulation materials during the test may result in material waste. Therefore, numerical simulation was used to model and analyze the test devices with different insulation thicknesses to select the best thickness of insulation material. As shown in Figure 4, the temperature cloud maps with and without an insulation layer (thickness 10 cm) at the age of 3 d, respectively, are shown. It is evident that without an insulation layer, the concrete test block loses heat rapidly and approaches room temperature by day 3. However, after adding an insulation layer, the temperature of the concrete test block dissipates slowly under the action of insulation layer. When the age is 3 d, the temperature of the concrete in the core area is 48 °C.

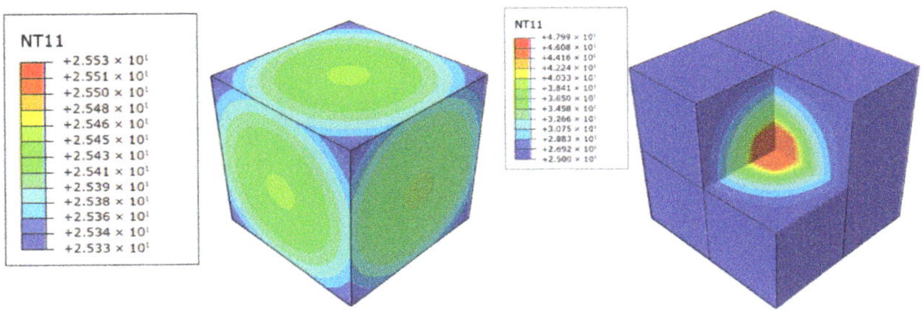

(**a**) No thermal insulation layer (**b**) With insulation layer

Figure 4. Temperature cloud image of semi-adiabatic temperature rise test device.

Furthermore, test setups with 20 cm and 30 cm thick insulation layers were numerically simulated. Figure 5 shows the core temperature curve of concrete under different thicknesses of the insulation layer, which shows that the core temperature significantly differs between concrete that has an insulation layer and concrete that does not. The highest temperature of the concrete without an insulation layer is only 35 °C, while the highest temperature after insulation layer is 56 °C. However, with the insulation layer thickness increasing from 10 cm to 30 cm, the temperature change trend of the test block is the same, and the size difference is not significant: thicker insulation layers do not always provide better effectiveness, and it will be more economical to choose a smaller thickness of the insulation layer based on a certain thickness. Therefore, a 10 cm thick insulation layer was selected for the semi-adiabatic temperature rise test in this study.

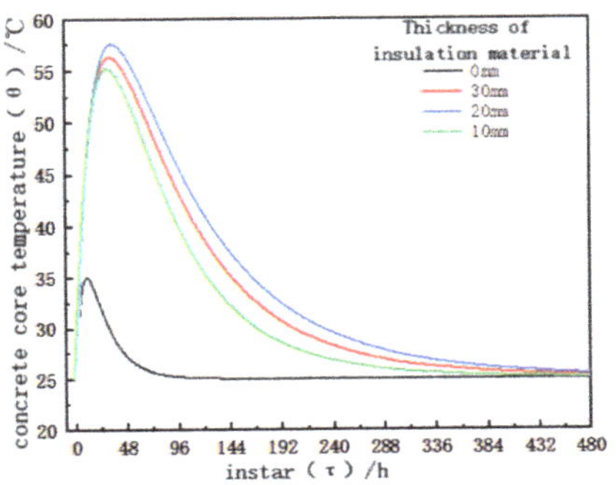

Figure 5. Core temperature curve of concrete test blocks with different insulation thicknesses.

3.2. Influence of Thermodynamic Parameters of Thermal Insulation Materials on Semi-Adiabatic Temperature Rise Test

At present, there are many kinds of insulation materials that can be applied, and the more commonly used ones are rigid foam polyurethane insulation board, extruded polystyrene board, rubber and plastic insulation board, etc. These insulation materials have good insulation effects, but their thermal properties are different. To study the effect of the specific heat capacity of insulation material on the semi-adiabatic temperature rise test, the specific heat capacity of the insulation layer was set as 1500 J/(kg·°C), 1200 J/(kg·°C), and 900 J/(kg·°C), respectively, in the model, and finite element analysis was performed to obtain the core temperature curves of concrete test blocks under various specific heat capacities of insulation layers, as shown in Figure 6. The figure indicates that the insulation material's specific heat capacity has minimal impact on the semi-adiabatic temperature rise test results, with the three temperature change curves almost coinciding. This is because in the semi-adiabatic temperature rise test device, the main function of the incubator is to insulate the entry of heat from the outside, and the specific heat capacity affects the heat storage capacity of the material, which minimally affects the temperature variation of the test block.

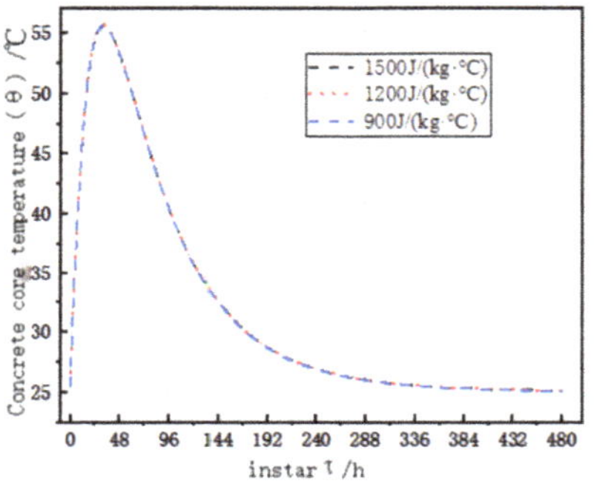

Figure 6. Core temperature curves of concrete test blocks with insulation layers with different specific heat capacities.

Similarly, to study the influence of thermal conductivity of thermal insulation materials on the semi-adiabatic temperature rise test, the thermal conductivity of thermal insulation layer was set as 0.024 W/m·K, 0.030 W/m·K, and 0.041 W/m·K, respectively, in the model, and finite element analysis was performed to obtain the core temperature curves of concrete test blocks under thermal insulation layers with different thermal conductivities, as shown in Figure 7. The figure shows that the thermal conductivities of various insulation materials are different, and the core temperature of the concrete test blocks is very different: the greater the thermal conductivity of the insulation material, the lower the core temperature of the concrete, and the shorter the time to reach the highest temperature. This is because the thermal conductivity of the thermal insulation material is directly related to its thermal insulation capacity; the smaller the thermal conductivity, the stronger the thermal insulation capacity, and the higher the temperature of the concrete test block, the more it can reflect the temperature change characteristics of concrete. It can be concluded that when the thermal conductivity of the insulation layer is less than 0.041 W/m·K, materials with good thermal insulation ability (low thermal conductivity) should be selected as much as possible.

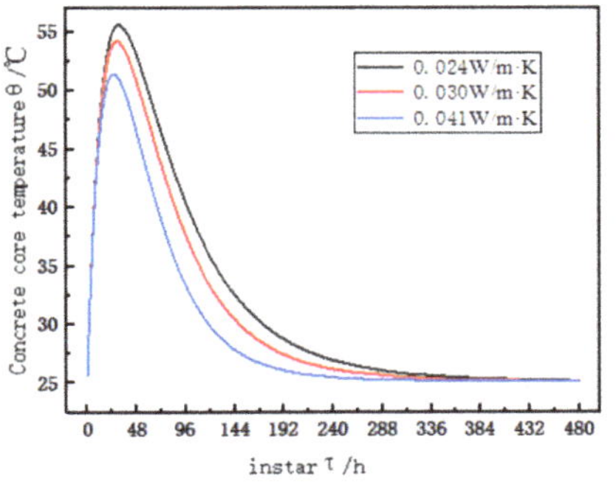

Figure 7. Core temperature curves of concrete test blocks under different thermal conductivity insulation layers.

3.3. Influence of the External Environment of the Insulation Layer on the Semi-Adiabatic Temperature Rise Test

In the semi-adiabatic temperature rise test device, the surface of the insulation layer directly contacts the external environment, so the roughness of the surface of the insulation material and the ambient wind speed may affect the heat loss rate between the device and the outside world. To investigate how the external environment of the thermal insulation layer affects the semi-adiabatic temperature rise test, the heat loss coefficient of the film layer on the thermal insulation surface was set as 10,000 W/(m^2·K), 14,000 W/(m^2·K), and 17,000 W/(m^2·K) in the finite element analysis to simulate the test under different external environments. Figure 8 shows the core temperature curves of concrete subjected to different external insulation conditions. When the heat dissipation coefficient of the coating is set between 10,000 and 17,000 W/(m^2·K), the heat dissipation coefficient of various insulation layers has minimal impact on the core temperature of the concrete test block, with the three curves nearly overlapping. This is because the heat insulation effect of the insulation material makes the external environment have less influence on the temperature change of the internal concrete test block, the specific heat capacity of the insulation material is large, and the change in external heat dissipation conditions has less influence on it. Therefore, the semi-adiabatic temperature rise test in this study did not carry out additional treatment on the surface of the insulation material, and the test site was arranged in the general concrete test site.

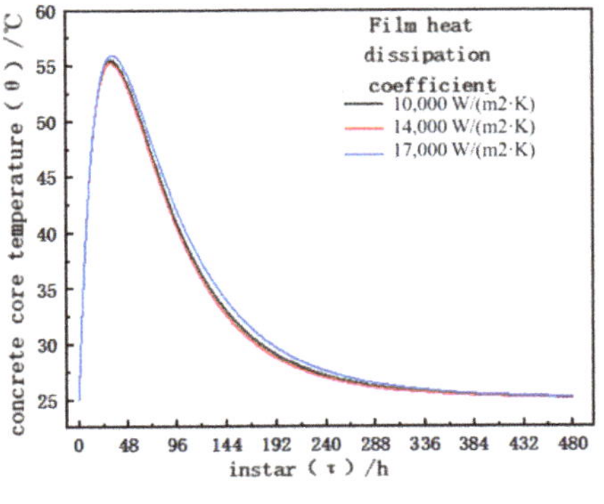

Figure 8. Core temperature curve of concrete test block under different external environments of insulation layers.

In summary, the optimal design of the semi-adiabatic temperature rise test in this study is as follows: the test object is set as a concrete cube with a side length of 20 cm, a cube incubator is used to completely wrap the internal concrete test block, the insulation material is selected as a polyurethane insulation board with a thickness of 10 cm, the surface of the incubator is not treated, and the test is carried out on a conventional concrete test ground.

4. Semi-Adiabatic Temperature Rise Test Scheme and Data Analysis

In practical engineering, the most common anti-crack and anti-seepage measures are to use cement with low hydration heat to reduce the temperature of concrete. The main purpose of incorporating mineral admixtures into concrete is to reduce the heat of hydration generated during the construction of bulk concrete, thereby reducing the problem of structural cracking. At the same time, combined with the actual situation of the project, a variety of prevention and control measures are comprehensively used in the three stages of construction design, structural maintenance, and structural forming, so as to realize the complementary advantages of multiple measures, which can reduce the occurrence of concrete cracks in the subway station with the greatest benefit, ensure the safety and quality of the structure, and reduce the construction cost. In recent years, with the continuous research on low-hydration thermal mineral admixtures, increasingly more mineral admixtures have been applied in mass concrete, among which the more common ones are fly ash, slag, limestone powder, metakaolin, etc. [24–26], which can not only reduce the hydration heat of concrete, but also improve various other properties of concrete. In this study, an optimized semi-adiabatic temperature rise test was carried out on several mineral admixture concretes, and the test data were processed and analyzed to summarize the temperature variation performance of different mineral admixture concretes. The models of hydration and heat release of concrete with different mineral admixtures were obtained using the finite element analysis method. The materials with the best temperature drop effect were obtained.

4.1. Semi-Adiabatic Temperature Rise Test Scheme

4.1.1. Test Device

According to the above numerical simulation optimization design of the semi-adiabatic temperature rise test, the semi-adiabatic test device mainly included two parts: a concrete mold and an incubator. The temperature in the incubator was kept at room temperature. The external concrete mold box was a cube with a side length of 31 cm, and it was composed of 10 cm thick polyurethane composite insulation board. The internal concrete mold box was a cube with a side length of 21 cm, composed of 1 cm thick polypropylene material sheet, and the internal space of the box was 20 cm × 20 cm × 20 cm. In this test, an embedded temperature sensor with an accuracy of ±0.25 °C was installed at the core of the concrete test block to measure the core temperature of the concrete test block. An automatic comprehensive testing system was employed to continuously monitor the temperature changes in the concrete, with measurements taken every 20 min, allowing for automatic storage of temperature data from the concrete test block. The test device and monitoring instrument are shown in Figure 9.

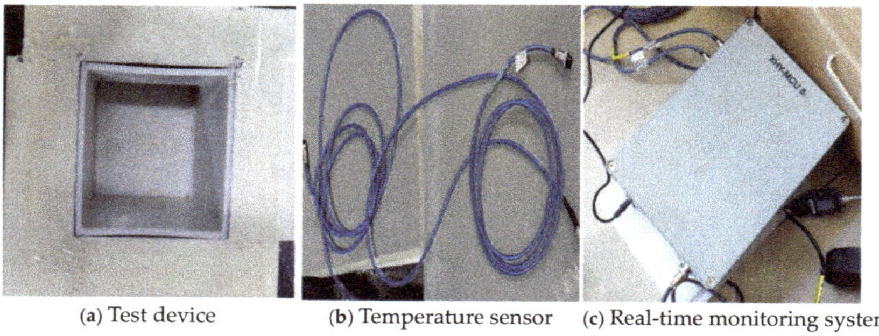

(**a**) Test device (**b**) Temperature sensor (**c**) Real-time monitoring system

Figure 9. Semi-adiabatic temperature rise test device and monitoring instrument [23].

4.1.2. Mix Ratio of Concrete Test Block

In mass concrete projects, mineral admixtures typically replace 20–70% of the cement [27]. Therefore, this test replaced 40% of the cement with different mineral admixtures to analyze their impact on the temperature rise properties of concrete. The standard C40 concrete mix ratio was used for the test block. Fly ash is a commonly used mineral admixture in engineering to lower the heat of hydration [28]. Therefore, this test used fly ash, and the other three materials (slag, limestone powder, metakaolin) were arranged and combined to divide the concrete test blocks into double-doped, triple-doped, and four-doped mineral admixtures, and semi-adiabatic temperature rise tests were conducted. The materials were labeled as follows: F (fly ash), S (slag), L (limestone powder), and K (metakaolin)—the numbers after the letters represent the percentage of the replaced cement. For example, F20S20L0K0 indicates the concrete test block mixed with 20% fly ash and 20% slag, and the specific combination of other test blocks is shown in Table 2.

Table 2. Mix ratio of multi-mineral admixture concrete [23].

ID	Cement(kg)	Fly ash (kg)	Slag (kg)	Limestone powder (kg)	Metakaolin (kg)	Sand (kg)
F0S0L0K0	420.0	0.0	0.0	0.0	0.0	745.0
F20S20L0K0	252.0	84.1	84.1	0.0	0.0	745.0
F20S0L20K0	252.0	84.1	0.0	84.1	0.0	745.0
F20S0L0K20	252.0	84.1	0.0	0.0	84.1	745.0
F15S12.5L12.5K0	252.0	63.0	52.5	52.5	0.0	745.0
F15S12.5L0K12.5	252.0	63.0	52.5	0.0	52.5	745.0
F15S0L12.5K12.5	252.0	63.0	0.0	52.5	52.5	745.0
F10S10L10K10	252.0	42.0	42.0	42.0	42.0	745.0

ID	Stone (kg)	Water (kg)	Admixture (kg)	Volume (m^3)
F0S0L0K0	1117.0	168.0	2.5	0.008
F20S20L0K0	1117.0	168.0	2.5	0.008
F20S0L20K0	1117.0	168.0	2.5	0.008
F20S0L0K20	1117.0	168.0	2.5	0.008
F15S12.5L12.5K0	1117.0	168.0	2.5	0.008
F15S12.5L0K12.5	1117.0	168.0	2.5	0.008
F15S0L12.5K12.5	1117.0	168.0	2.5	0.008
F10S10L10K10	1117.0	168.0	2.5	0.008

4.1.3. Test Process

The semi-adiabatic temperature rise test was divided into the following six steps. (1) Assembly of the customized polyurethane insulation board insulation box and polypropylene concrete mold. (2) The test materials were accurately measured according to the mix ratios of the different concrete test blocks and were fully mixed according to the test standards. (3) The mixed concrete was added to a polypropylene concrete mold and fully vibrated. (4) A temperature sensor was embedded in the core of the concrete test block at its reserved position. (5) The concrete test block was placed together with the mold in the preset incubator and sealed. (6) The real-time monitoring system was accessed to extract temperature data every 20 min, monitor the concrete test block for about seven days, and finally obtain the semi-adiabatic temperature rise curve of the concrete test block during the age. Part of the test site pictures are shown in Figure 10.

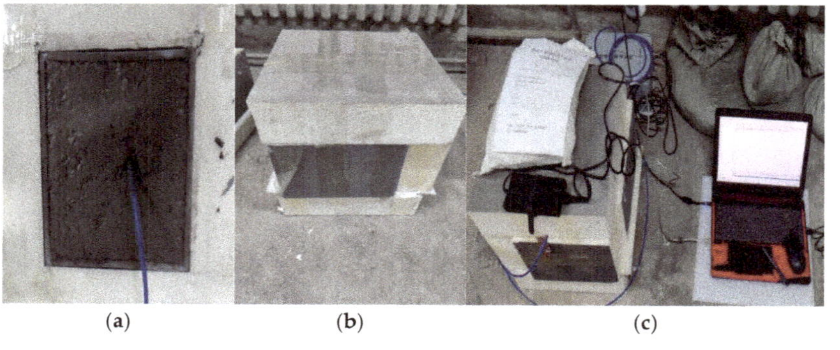

Figure 10. Field trial flow chat. (a) Embedded temperature Sensor. (b) Insulation material used to seal concrete. (c) Access to real-time temperature monitoring system.

4.2. Analysis of Test Data

4.2.1. Temperature Change Data Analysis

After processing the test data, the rate of temperature rise curve of the concrete test block mixed with two types of mineral admixtures and a typical concrete test specimen was obtained, as shown in Figure 11. The following can be seen from Figure 11a: (1) after adding two types of mineral admixtures, the maximum temperature of the concrete core is significantly reduced, which suggests that various combinations of two mineral admixtures can effectively lower the temperature of concrete as it ages. (2) Different double-doped mineral admixture combinations impact the temperature of concrete in distinct ways, in which F20S0L20K0 has the slowest temperature rise in the early stage and reaches the highest temperature of 41.75 °C, but the temperature drop is slower in the cooling stage. In the early stage, the temperature rise of F20S20L0K0 responds more rapidly, and the highest temperature is 44.75 °C, but the temperature drop is the slowest in the cooling stage. The temperature rise of F20S0L0K20 is the fastest in the early stage, and the highest temperature reached is 46.25 °C, but the temperature drop is the fastest in the cooling stage. The concrete test block with double fly ash and limestone powder exhibits the best cooling effect. Figure 11b illustrates that incorporating three mineral admixtures can also help lower the temperature of concrete during its age. The temperature rise of F15S12.5L12.5K0 is slow in the early stage, the maximum temperature is 44 °C, and the temperature drop is slow in the cooling stage. The temperature of F15S0L12.5K12.5 increased rapidly in the early stage, the highest temperature reached was 46 °C, and the temperature decreased rapidly in the cooling stage. The temperature curve of F15S12.5L0K12.5 is like that of F15S0L12.5K12.5. The temperature of F15S12.5L12.5K12.5 is higher than that of F15S0L12.5K12.5. The temperature of F15S12.5L12.5K12.5 is higher than that of F15S0L12.5K12.5.

In summary, the cooling effect of the concrete with three kinds of three-doped mineral admixtures is worse than that of the concrete with three kinds of double-doped mineral admixtures, and F15S12.5L12.5K0 exhibits the best cooling effect. Figure 11c shows that four-doped mineral admixtures also have a certain effect on reducing the temperature of concrete. The maximum temperature that F10S10L10K10 can achieve is 44.75 °C, and its effect is between that of double-doped and triple-doped mineral admixtures. This shows that the content and performance of different mineral admixtures affect the temperature change of concrete, which has little relationship with the type of mineral admixtures.

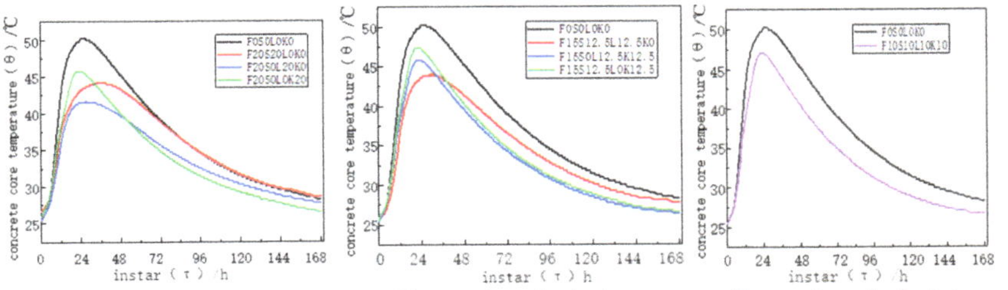

(a) Two-mineral admixtures (b) Three-mineral admixtures (c) Four-mineral admixtures

Figure 11. Temperature curves of concrete test blocks with different mineral admixtures.

4.2.2. Temperature Fitting Curve Analysis

Fitting analysis of the temperature curves obtained from these experiments was performed. In this study, polynomial fitting was used to fit the warming and cooling sections of the temperature curve, so that the R2 of the fitting curve reached 0.99. To ensure the fitting degree of the fitting curve, as shown in Figure 12, the temperature fitting curve of concrete mixed with two types of mineral admixtures was quite different, but its heating stage was mainly around 24 h. In the heating stage, F0S0L0K0 had the fastest temperature rise and the highest temperature, the concrete test block mixed with limestone powder had the slowest temperature rise and the lowest temperature reach, and the concrete test block mixed with slag and metakaolin was in the middle. In the cooling section, F0S0L0K0 had the fastest cooling rate, the concrete test block mixed with metakaolin had the second highest cooling rate, and the concrete test block mixed with slag and limestone powder had the slowest cooling rate, which shows that the performance of limestone powder was the best and that of metakaolin was the worst for reducing the heat of hydration and the speed of the hydration reaction.

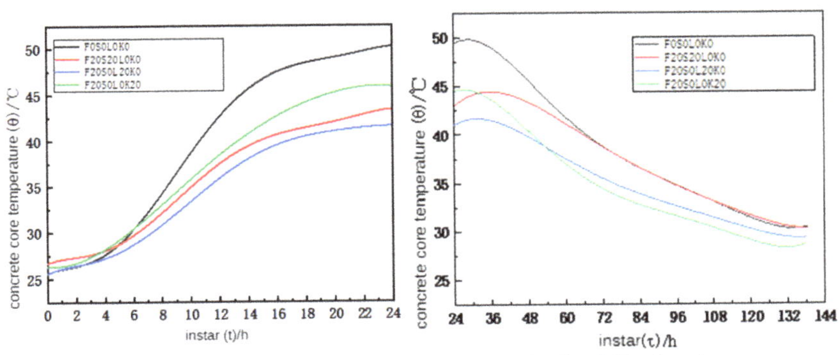

(a) before 24 h of ageing (b) after 24 h of ageing

Figure 12. Temperature fitting curve of concrete test block with two types of mineral admixtures.

Similarly, the rate of temperature increase curves of the concrete test block with three and four mineral admixtures were fitted, as shown in Figure 13. The temperature-fitting curves of concrete mixed with more than three types of mineral admixtures are not significantly different from those mixed with two types of mineral admixtures. In contrast, the quantity and variation trend of these temperature fitting curves are similar, which indicates that the temperature rise performance of the concrete test specimens tends to be the same when the types of mineral admixtures are increased. The temperature fitting curve with the best temperature control effect was F15S12.5L12.5K0, which further indicates that the cooling performance of limestone powder and slag is better than that of metakaolin.

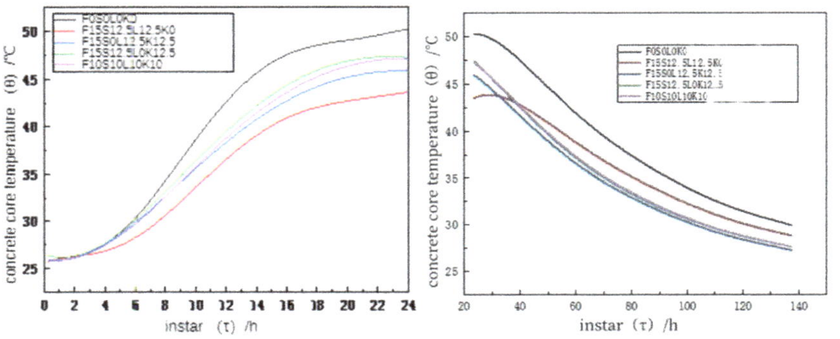

(**a**) before 24 h of ageing (**b**) after 24 h of ageing

Figure 13. Temperature fitting curve of concrete test block mixed with three or four mineral admixtures.

For the cement hydration reaction, the temperature change rate usually goes through five periods: initial, induction, acceleration, deceleration, and decay [29]. To further analyze the temperature variation characteristics of the multi-doped mineral admixture concrete, the temperature fitting curve equation above was applied to obtain the rate of temperature increase curve of the double-doped mineral admixture concrete test block, as shown in Figure 14. The temperature change of concrete has obviously experienced the induction period, acceleration period, deceleration period and decay period, which is consistent with Jia Fujie's conclusion [19], and the initial period cannot be reflected in the figure because it is in the process of concrete preparation. The rate of temperature increase for F0S0L0K0 in the acceleration period is much higher than that of other mineral admixture concrete, with a maximum temperature rise rate of 2.2 °C/h. However, the rate of temperature increase for F0S0L0K0 in the deceleration period decreases rapidly and is generally lower than that of other concrete in the decay period. This indicates that ordinary Portland cement has a large heat release and a severe hydration reaction, and adding mineral admixtures helps to reduce the amount of hydration heat released and the rate of the hydration reaction, which contributes to better control over the temperature of concrete. Among them, F20S20L0K0 and F20S0L20K0 have the smallest temperature change rate and the smallest temperature change range, and the maximum temperature rise rate is 1.25 °C/h. Therefore, to reduce the temperature of mass concrete, slag and limestone powder, two mineral admixtures, are preferred to be used with fly ash.

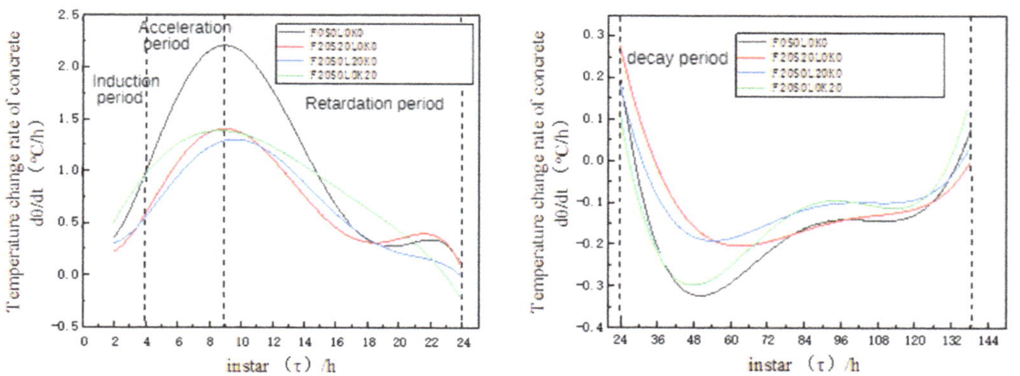

(**a**) before 24 h of ageing (**b**) after 24 h of ageing

Figure 14. Temperature rise rate curve of concrete test block with two types of mineral admixtures.

Similarly, the rate of temperature increase curve of the concrete test block with more than three mineral admixtures was obtained by derivation, as shown in Figure 15. Compared with the curve of F0S0L0K0, the temperature rise rate curves of several multi-mineral admixture concrete test blocks show little difference, among which the temperature rise rate of F15S12.5L12.5K0 is small, and the change range is small, which is the same as the above conclusion. Therefore, in the concrete mixed with more than three kinds of mineral admixture, the temperature rise rate of F15S12.5L12.5K0 is small. Therefore, the three mineral admixtures, namely, fly ash, limestone powder, and slag, are preferred.

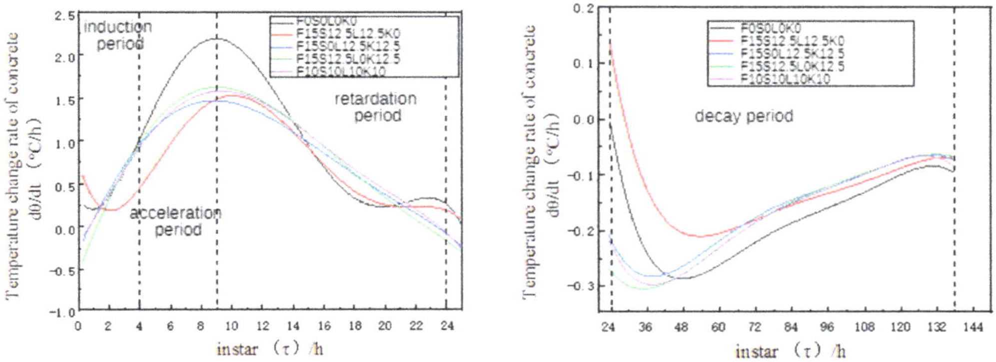

(**a**) before 24 h of ageing (**b**) after 24 h of ageing

Figure 15. Temperature rise rate curve of concrete test block with three or four kinds of mineral admixtures.

In summary, F20S0L20K0 exhibits the best performance in reducing the temperature of concrete. To better compare and analyze the temperature rise performance of concrete with different mineral admixtures, it was analyzed from the following three dimensions: (1) the highest temperature that the concrete test block can reach; (2) the age when the concrete test block reaches the highest temperature; (3) the temperature of the concrete test block at the age of 5 d. These three can reflect the total cooling capacity, preheating rate, and mid–late cooling rate of the concrete, respectively.

As shown in Table 3, F20S0L20K0 had the lowest maximum temperature and maximum temperature rise rate. Compared with the other test blocks, limestone powder was added to F20S0L20K0, indicating that limestone powder had the most obvious cooling effect and the smallest total heat release. F20S0L0K20 takes the shortest time to reach the highest temperature. Compared to the other test blocks, metakaolin was added to F20S0L0K20, which indicated that metakaolin had the fastest heat generation rate in the early stage. F20S20L0K0 had the highest temperature on the fifth day. Compared with other test blocks, slag was added to F20S20L0K0, indicating that the hydration reaction rate of slag was slower, and the heat generation rate was the fastest in the later stage. Through an analogy analysis of the temperature variation data of three and four blends, the following rules can be obtained: (1) total cooling capacity: limestone powder > fly ash > slag > metakaolin; (2) early heat generation rate: metakaolin > slag > fly ash > limestone powder; (3) heat reduction rate in the middle and late period: metakaolin > limestone powder > fly ash > slag. Replacing the same proportion of cement does not have a better effect than adding more types of mineral admixtures, and the most suitable ingredients must be found according to the characteristics of each mineral admixture.

Table 3. Temperature rise performance of concrete with different mineral admixtures.

Specimen Number	Maximum Temperature (°C)	Age at Which the Maximum Temperature Is Reached (h)	Temperature at Age 5 d (°C)	Maximum Temperature Rise Rate (°C/h)
F0S0L0K0	50.25	24.00	31.50	2.25
F20S20L0K0	44.25	38.00	31.25	1.35
F20S0L20K0	41.75	27.00	30.25	1.25
F20S0L0K20	45.75	22.00	29.00	1.36
F15S12.5L12.5K0F	43.00	30.00	30.00	1.40
15S0L12.5K12.5F1	45.75	22.00	28.75	1.45
5S12.5L0K12.5F10	47.25	21.00	28.50	1.60
S10L10K10	47.00	23.00	28.00	1.50

5. Conclusions

To study the temperature rise characteristics of concrete with different mineral admixtures, a semi-adiabatic temperature rise test was conducted based on the principles of concrete hydration and heat release. A numerical simulation method and an optimized semi-adiabatic temperature rise test method were proposed. The following conclusions are drawn from the analysis of the test data:

(1) The optimal design of the semi-adiabatic temperature rise test was as follows. The test object was set as a concrete cube block with a side length of 20 cm, a cube incubator was used to completely wrap the internal concrete test block, the optimal thickness of insulation material is 10 cm, the insulation material is a polyurethane insulation material with good thermal insulation ability (low thermal conductivity), the change in external heat dissipation conditions has less influence on temperate rise so the surface of the incubator was not treated, and the test was carried out on a conventional concrete test ground.

(2) For mineral admixture concrete, the reduction in hydration heat is independent on the type of mineral admixture added. The decrease in hydration heat is related to the content of different mineral admixtures and material properties. An optimized semi-adiabatic temperature rise test was conducted, and the temperature rise performance of four common mineral admixtures was compared as follows: (1) total cooling capacity: limestone powder > slag, fly ash > metakaolin; (2) early heat generation rate: metakaolin > slag > fly ash > limestone powder; (3) heat reduction rate in the middle and late periods: metakaolin > limestone powder > fly ash > slag. Instead of the same content of cement, the concrete mixed with fly ash and limestone powder exhibited the best cooling performance, and it is recommended for use in engineering.

In addition, there are still shortcomings in this paper, which can be improved from the following two points:

(1) Many experiments are needed to fit the data of the semi-adiabatic temperature rise curve to provide reliable theoretical support for the project.

(2) Try to make a simple semi-adiabatic temperature rise test model that can simulate the on-site structure to avoid the influence of size on the test.

(3) The finite element simulation part of this paper is to provide reasonable parameter settings for the experimental part, since the content of the simulation part is not consistent with the experimental part. This is an inadequacy of this article. Therefore, the experimental content can be simulated for further validation.

Author Contributions: Conceptualization, K.W.; methodology, K.W., C.W. and J.X.; software, Z.L. and J.X.; validation, Z.L. and T.Y.; formal analysis, K.W. and Z.L.; investigation, Z.D., C.W. and T.Y.; resources, C.W. and T.Y.; data curation, Z.D. and J.X.; writing—original draft preparation, K.W.; writing—review and editing, Z.L.; visualization, K.W.; supervision, K.W.; project administration, K.W.; funding acquisition, K.W. All authors have read and agreed to the published version of the manuscript.

Funding: This research received no external funding.

Data Availability Statement: Some or all data, models, or codes that support the findings of this study are available from the corresponding author upon reasonable request.

Conflicts of Interest: Authors C.W. and T.Y. were employed by the company Power China Municipal Construction Group Co., Ltd. The remaining authors declare that the research was conducted in the absence of any commercial or financial relationships that could be construed as a potential conflict of interest.

References

1. Action Plan for Green Urban rail Development of Urban rail Transit in China. *China Metros* **2022**, *8*, 20–35.
2. Yuan, G.L.; Huang, F.Y.; Shen, H.; Gao, P.F. Research on temperature field and thermal stress of hydration heat in massive concrete of construction period. *Concrete* **2005**, *2*, 86–88.
3. Qin, W.Z. Shrinkage and cracking of concrete and its evaluation and prevention. *Concrete* **2001**, *7*, 3–7.
4. Lin, Z.H.; Qin, W.Z.; Zhang, S.H.; Zhang, T. Development of stress and evaluation of cracking sensitivity of concrete at early ages. *Archite Technol.* **2003**, *1*, 34–35.
5. Zhou, Z.P. Water leakage control in the main structure of a metro station. *Mod. Tunn. Technol.* **2012**, *49*, 164–170.
6. Duan, Z.; Ma, C.L.; Wang, L.L. Causes and prevention of cracks in cast-in-place mass concrete. *Concrete* **2003**, *5*, 48–58.
7. Zhu, Y.T.; Zhan, S.L. Study on reasons and prevention measures of cracking in concrete tructure. *J. Mater. Sci. Eng.* **2003**, *5*, 727–730.
8. Jiang, X.P.; Wang, S.L.; Duan, S.X.; Sun, Y. Mechanism analysis of temperature crack appear of super mass concrete and new methods of anti-rack. *Concrete* **2007**, *12*, 98–102.
9. Chen, Y.; Zhang, Q.S. Review of study on cement concrete cracking resistance at early ages. *J. Build. Mater.* **2004**, *4*, 411–417.
10. Aydin, B.B.; Tuncay, K.; Binici, B. Overlapping Lattice Modeling for concrete fracture simulations using sequentially linear analysis. *Struct. Concrete* **2018**, *19*, 568–581. [CrossRef]
11. Gallo, P.; Hagiwara, Y.; Shimada, T.; Kitamura, T. Strain energy density approach for brittle fracture from nano to macroscale and breakdown of continuum theory. *Theor. Appl. Fract. Mech.* **2019**, *103*, 102300. [CrossRef]
12. Xu, S.L.; Dong, L.X.; Wang, B.W.; Li, Q.H. Development of concrete fracture mechanics in China during the past three decades. *J. Hydraul. Eng.* **2014**, *45*, 1–9.
13. Zhang, X.F.; Xu, S.L. Determination of fracture energy of three-point bending concrete beamusing relationship between load and crack-mouth opening displacemen. *J. Hydraul. Eng.* **2008**, *6*, 714–719.
14. Xu, S.L.; Bu, D.; Zhang, X.F. A study on double-K fracture parameters by using wedge-splitting test on compact tension specimens of various sizes. *China Civ. Eng. J.* **2008**, *2*, 70–76.
15. Guo, Q.J. Experimental Investigation on the Influence of Coarse Aggregate for Mortar Crack Propagation. Master's Thesis, Dalian University of Technology, Liaoning, China, 2016.
16. Bažant, Z.P.; Oh, B.H. Crack band theory for fracture of concrete. *Mater. Constr.* **1983**, *16*, 155–177. [CrossRef]
17. Zhan, J.; Liu, Q.; Wang, L. Conditions promoting crack growth in concrete along the aggregate/matrix interface or into the aggregate. *J. Tsinghua Univ. Sci. Technol.* **2004**, *3*, 387–390.
18. Wang, L.C.; Wu, D.; Bao, J.W.; Liang, Y.J. Mesoscale numerical simulation analysis for temperature distribution in early-age concrete. *J. Hydraul. Eng.* **2017**, *48*, 1015–1022.
19. Jia, F.J. Research on Concrete Half-Adiabatic Test and Finite Element Calculation. Master's Thesis, China Building Materials Academy, Beijing, China, 2011.
20. Yun, H.D.; Park, W.S.; Jang, Y.I.; Kim, S.W. Prediction of the heat of hydration of fly ash concrete by adiabatic temperature rise test and regression analysis. *Mag. Concrete Res.* 2024, *in press*. [CrossRef]
21. Liang, T.; Luo, P.; Mao, Z.; Huang, X.; Deng, M.; Tang, M. Effect of Hydration Temperature Rise Inhibitor on the Temperature Rise of Concrete and Its Mechanism. *Materials* **2023**, *16*, 2992. [CrossRef]
22. Quan, J.J.; Fu, S.J.; Yang, R.D.; Chen, J.; Zhang, K.F. Characteristics and Modelling of Adiabatic Temperature Rise Rate of Medium-volumed Fly-Ash Concrete. *J. Sci. Res. Inst.* **2023**, *40*, 134.
23. Wu, K.; Dou, Z.Y.; Liu, Z.H.; Xu, J.X. Study on Hydration Heat Release Model and Its Influence Coefficient of Addition Concrete. *Appl. Sci.* **2024**, *14*, 2276. [CrossRef]
24. Han, H.W. The influence of mineral admixtures on the hydration heat of cementitious materials. *Low Temp. Archit. Technol.* **2020**, *42*, 37–40.
25. Zeng, J.J.; Wang, S.N.; Xiong, J.B.; Liu, H.X. Influence of metakaolin on hydration performance of low heat Portland cement. *Concrete* **2020**, *5*, 69–72.
26. Liu, S.H.; Yan, P.Y. Hydration properties of limestone powder in complex binding material. *J. Chin. Ceram. Soc.* **2008**, *10*, 1401–1405.
27. *DL/T5055-2007*; Technical Specification of Fly Ash for Use in Hydraulic Concrete. National Development and Reform Commission: Beijing, China, 2007.

28. Zhang, Y. Study on the Volume Stability of Limestone Powder-Slag-Fly Ash Concrete and Hydration Heat of Composite Cementitious Materials. Master's Thesis, China University of Mining and Technology, Xuzhou, China, 2018.
29. Chen, Z.J.; Lei, D.; Hong, M. Study on temperature distribution and cracking of early-age concrete in semi-adiabatic state. *Henan Sci.* **2019**, *40*, 390–395.

buildings

MDPI

Article

Study on the Optimized Perception of Structural Behavior in Shield Tunneling by Fiber Grating Layouts

Rongjun Xing [1,2], Yufu Li [1,2], Chuan He [3], Daiqiang Zhu [1,2], Yujun Li [1,2], Kun Feng [3], Pai Xu [1,2,*], Yu Liu [1,2] and Zhongchao Zhao [1,2]

1 State Key Laboratory of Mountain Bridge and Tunnel Engineering, Chongqing Jiaotong University, No. 66, Xuefu Avenue, Nan'an District, Chongqing 400074, China; xingrong-12@163.com (R.X.); 18284886470@163.com (Y.L.); zhu_daiqiang@163.com (D.Z.); m15730446752@163.com (Y.L.); 15215150872@163.com (Y.L.); zc-zhao@mails.cqjtu.edu.cn (Z.Z.)

2 School of Civil Engineering, Chongqing Jiaotong University, Chongqing 400074, China

3 Key Laboratory of Transportation Tunnel Engineering, Ministry of Education, Southwest Jiaotong University, Chengdu 610031, China; chuanhe21@163.com (C.H.); windfeng813@163.com (K.F.)

* Correspondence: xu_pai@126.com

Abstract: Shield tunnels' structural stability is challenged due to the fact that they are often built under rivers, lakes, and oceans. It is crucial to execute the structural deformation perception of the shield tunnel. Fiber Bragg grating (FBG) sensing technology is sensitive to deformation information, making it one of the greatest options for shield tunnels to perceive structural deformation. In this study, a 1:20 scale model test was carried out to investigate the deformation perception of the shield tunnel structure under three different layouts of surface-mounted FBG sensors. The deformation law of the tunnel is discussed, under the condition of two-factor cross fusion and especially under the condition of constant water pressure and soil pressure change. The results indicate that, under the combined action of water and soil pressure, the uniform water pressure of 0.33 MPa has a stabilizing effect on the segment strain under the vertical load of 0.4 MPa. The traditional four-point layout and the 18° uniform layout are more effective in detecting changes in local tunnel curvature and strain, respectively, compared to the 36° uniform layout mode. It is advised that the traditional four-point layout be used to collect information for other sections' monitoring and that the 18° uniform layout is for harsh terrain conditions.

Keywords: shield tunnel; segment lining; model test; fiber Bragg grating; deformation perception

1. Introduction

Shield tunnel construction is fraught with terrain factors like high water pressure, high ground loads, and fault fracture zones. These factors challenge the shield tunnel's structural safety. The water–earth pressure is the most frequent factor of tunnel operation, which leads to segment transverse convergence and deformation. As a result, it is essential to study the perception of structural deformation. Specifically, the key is to master the structure perception method and summarize the structural deformation rule.

The choice of monitoring transducer is the foundation of structural deformation perception. There are faults in traditional monitoring transducers [1], such as them having a single data type, low survival rate, poor accuracy, etc. Traditional monitoring methods cannot provide a realistic reflection of the lining structure's overall deformation [2]. With the rapid development and use of fiber Bragg grating (FBG) sensing technology in recent years, FBG sensing technology offers a wide range of potential applications in monitoring tunnel structure health [3–5].

Many researchers have conducted a lot of research on the use of FBG for tunnel monitoring. Sun et al. [6] determined the link between the wavelength change in the fiber Bragg grating (FBG) sensor and the stress of the anchor bolt and showed that the FBG sensor satisfies the monitoring needs of the tunnel structure. Using the FBG sensor for high-precision structure monitoring, Li et al. [7] confirmed that the sensor is appropriate for displacement and stress monitoring in quasi-static structure studies. Structural health monitoring (SHM) applications for tunnel roads have to be researched and developed by means of fiber Bragg grating (FBG) optical temperatures and strain sensors [8]. The findings in this research illustrate that the data gathered from embedded FBG optical sensors provide the essential information on how the pavement structure could sustain the load and information about the traffic intensity on the specific road section; the structural life of the pavement could be evaluated and predicted. Wang et al. [9] employed FBG sensors to continuously track the convergence deformation of tunnels. The FBG sensor could accurately assess the convergence deformation and satisfy tunnel management requirements, according to experiments. To track the three-dimensional stress of the rock surrounding the tunnel in the Liang et al. [10] investigation, FBG sensing technology is employed. The experimental results demonstrate that the relative error of the recorded stress is less than 4%, demonstrating that the FBG sensor has good linearity and repeatability and has a wide range of application prospects in the area of monitoring the safety of underground engineering structures such as coal mines and tunnels. A mechanical analysis model of shear strain transmission that takes into account the viscoelastic impact of optical fiber sensing apparatus was developed by Liang et al. [11]. The effect of shear modulus and cement layer thickness on quasi-static strain transfer was examined by the uniaxial tensile test of a bolt. Sigurdardottir et al. [12] found a substantial correlation between curvature change and structural deformation. According to research by Urszula et al. [13], sensing modal strain and curvature with an FBG sensor could increase monitoring sensitivity by around 30 times and be utilized to be more sensitive to local deformation or even damage. In experiments, Liu et al. [14] concurrently measured temperature and curvature using an integrated fiber optic sensor. The potential of FBG sensing components for structural deformation is demonstrated by the relative errors of temperature and curvature, which are, respectively, 2.22% and 2.58%.

Many researchers have conducted a lot of research on structure health monitoring. Sun et al. [15] proposed a sensor anomaly diagnosis method based on the characteristics that the original output of the fiber grating strain gauge has the same law in the same kind of structure monitoring. Zhou et al.'s team [16] used a three-dimensional point cloud as the data basis and used the fusion of machine vision structural disease detection information to monitor the health and safety status of the tunnel structure. Chen et al. [17] proposed a new comprehensive monitoring method for tunnel section deformation based on the inclination angle of monitoring points and further verified the accuracy of the method in practical applications through experiments. Zhou et al. [18] proposed a tunnel risk mathematical model and tunnel deformation monitoring method based on the three indicators of current change, cumulative change, and change rate of monitoring points. Sui et al. [19] arranged the distributed optical fiber along the circumferential direction on the upper and lower surfaces of the steel cage inside the segment and obtained the strain distribution inside the segment. Zheng [20] collected the curvature data of discrete points on the inner surface of the tunnel through neural network processing. Finally, a more accurate tunnel section curve is obtained. Letsios et al. [21] detects newly built buildings and estimates their height in the study area based on permanent scatterer interferometry (PSI) technology. Its overall goal is to obtain timely spatial information for construction activities. Using Permanent Scatterer Interferometry (PSI) analysis, Roccheggiani et al. [22] revealed regional subsidence in great detail, particularly where the ground is characterized by clay and alluvial deposits and where large buildings exist. Xu et al. [23] proposed a structural health monitoring method combining optical fiber sensing and piezoelectric sensing and constructed a structural load identification method based on optical fiber sensing. Ghaderpour et al. [24] uses the sequential turning point detection (STPD) method to estimate the trend turning point and its direction in the PS-InSAR time series to explore its potential impact on ground deformation.

Although numerous researchers have conducted pertinent research, the study of the shield tunnel deformation law is still in its early stage. Utilizing a three-dimensional soil-shield tunnel similar model test system, Tang et al. [25], and colleagues conducted experimental research on the structural deformation state and mechanical behavior of shield tunnel segment lining of Nanjing Metro station under various assembly modes. In a related model test for the Nanjing Yangtze River Tunnel, He et al.'s team [26] took into account the interactions between formation, water pressure, and tunnel segments and analyzed the impact of segment lining assembly mode on the internal force distribution of the structure. Kang et al. [27] studied the longitudinal deformation and damage of shield tunnels caused by ground stress and surface overload. It is proved that when the surface load is far away from the tunnel, the shallow shield tunnel has a large settlement, and it is concluded that increasing the equivalent shear stiffness will significantly reduce the dislocation between the longitudinal adjacent segment rings. Ding et al. [28] analyzed the effect of soft soil grouting on the deformation of the closed shield tunnel with the measured data. Combining the measured data of vertical, horizontal, and convergence deformation of the adjacent tunnel during the grouting construction in foundation pit engineering, the influence of grouting on the metro tunnel in soft soil area is analyzed. Zhang et al. [29] measured the influence of relevant tunnel coefficients on lining displacement and internal force through parameter analysis, that is, the soil Young's modulus, soil unit weight, soil lateral pressure coefficient, tunnel radius, tunnel depth, and gap parameters. Wang et al. [30] studied the progressive instability and failure process of the large-diameter underwater shield tunnel segmental lining structure, from the microscopic damage of the material to the overall instability of the structure, by the model test method. The progressive failure process is divided into three stages: initial elastic phase, local damage phase, and overall instability failure phase. And the influences of the assembly method, the position of the capping block, the lateral pressure coefficient of the stratum and the amount of water pressure on the deformation characteristics and the critical instability state of the segmental lining structure are, respectively, analyzed. To direct the construction design and carry out the targeted treatment of structure characteristics at the intersection of cross passages, Tan et al. [31] examined the structural deformation rule of large-diameter shield tunnel segments at the intersection of cross passages. Civera et al. [32] studied the seismic response of an underwater railway tunnel and two nearby highway tunnels in the same near-field seismic event. Arias Intensity (AI) and significant duration (Ds_{595}) are proposed as potential explanations for the different seismic susceptibility of these different tunnels.

The existing research results show that sensing information in tunnel health monitoring is single, and the correlation between indicators is not clear. FBG monitoring technology can effectively and accurately monitor the deformation data of curvature and strain. However, the existing research rarely discusses the influence of different FBG layouts on the accurate monitoring of the local strain and curvature of tunnels. Moreover, under the condition of two-factor cross-fusion, especially under constant water pressure and variable earth pressure, the structural deformation law is uncertain. Therefore, it is necessary to carry out relevant research. In this study, the strain and curvature of the segment's inner surface were sensed and measured using the surface-mounted FBG sensing element. The deformation law of the shield tunnel segment structure under the action of water–earth pressure is explored. The second section of this article mainly describes the basic situation and preparation of the test. The third section is the analysis and discussion of the test results. The main contents of the four sections are conclusions and deficiencies. The following contributions are provided: (1) By optimizing and strengthening the layout of structural monitoring, accurate deformation data can be provided and the requirements of structural health monitoring of shield tunnels can be met. (2) By collecting the deformation information of the shield tunnel segment under the combined action of water and soil pressure, the deformation characteristics of the segment are summarized. This provides a reference for health monitoring, disease prediction, and disaster warning threshold determination for shield tunnel segments.

2. Experimental Preparation

2.1. Project Description

Based on the Yellow River tunnel project, this paper carries out a scale test on the shield tunnel. The landform of the tunnel site is alluvial plain, and the terrain is relatively flat. The shield tunnel's smallest thickness of covering soil is 11.2 m, and its maximum thickness of covering soil is 42.3 m. The thickness of covering soil of the Yellow River section is 25 to 38 m. In this project, the Yellow River's flood level is 34.18 m, and the tunnel's maximum water pressure is roughly 0.65 MPa. The shield tunnel has two layers: a lower subway layer and an upper driving layer with three lanes. The lining ring is 15.2 m in outside diameter, 13.9 m in inner diameter, and 0.65 m thick. The material is reinforced concrete type C60 (cube compressive strength). The segment ring is 2.0 m wide and it is divided into 7 + 2 + 1 pieces. The standard section layout of shield tunnel is shown in Figure 1.

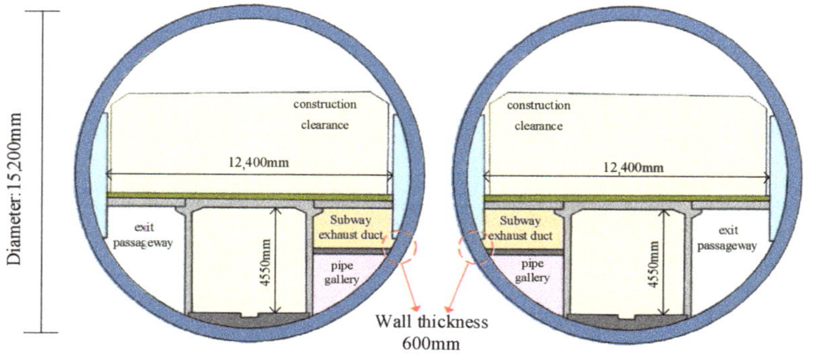

Figure 1. Standard cross-section layout of shield tunnel.

2.2. Soil Materials and Lining Models

The similarity theory in this section mainly refers to the second similarity theorem, also known as the π theorem. According to the similarity criterion corresponding to the similar mechanical phenomena, Equation (1) can be obtained. We select a basis similarity ratio of 1:20 for geometrical similarity and 1:1 for bulk density similarity to regulate the similarity of physical and mechanical characteristics in the elastic range. According to the basis relation, the similarity relation of strain, friction force, and Poisson's ratio is $C_\mu = C_\varepsilon = C_\varphi = 1$. The similarity relation of strength, stress, cohesion, and elastic modulus is $C_R = C_s = C_c = C_E = 20$. The heavy similarity ratio in terms of geometry is $C_\gamma = \gamma_p/\gamma_m = 1$.

$$\frac{C_\sigma}{C_X C_l} = 1;\ C_\varepsilon = 1;\ \frac{C_E}{C_X C_l} = 1;\ C_\mu = 1;\ \frac{C_{\overline{X}}}{C_X C_l} = 1;\ \frac{C_\delta}{C_l} = 1 \tag{1}$$

The ratio of soil mass–similar materials is based on the test geometric similarity ratio of 1:20, and the values of soil–similar materials are based on the project's real physical and mechanical properties. The mechanical parameters are shown in Table 1. According to the soil parameters, combined with the study of the shield similar stratum model [33], a direct shear test and compression test were conducted. Then, the weights of barite powder, fly ash, oil, and river sand were determined. The weight ratio is shown in Table 2. The similar parameters of model test segment are shown in Table 3.

Table 1. Test soil similarity parameters.

Mechanical Parameters	Severe γ	Elastic Modulus	Cohesive Forces	Internal Friction Angle
	(kN/m^3)	(MPa)	(MPa)	φ (°)
Prototype soil	19	45	38	20
Model soil	19	2.25	1.95	20

Table 2. Test soil material ratio (weight ratio).

Barite	Fly Ash	Engine Oil	Sand
0.25	0.225	0.125	0.4

Table 3. Similar parameters of model test segment.

Mechanical Parameters	Gravity	Elastic Modulus	Single Axis Compressive Strength	Poisson's Ratio
	(kN/m^3)	(GPa)	(MPa)	
Prototype segment	23	36	38.5	0.2
Mode segment	23	1.8	1.925	

The segment model was made according to the similarity ratio, and the similar materials were made according to the ratio of fixed water–gypsum = 1:1.4. The shield tunnel segment joint was simulated to understand the special link between the segment and the ring indirect head of the shield tunnel. For the joint form of the longitudinal joint of the segment, the longitudinal joint is simulated by slotting. A certain depth of slot is cut in the model ring corresponding to the part of the tunnel joint to weaken the bending stiffness of the part. The slot depth is determined by the equivalent bending stiffness of the prototype joint [34]. To join the segment ring to the ring and simulate the indirect head of the segment ring, we used a bolt with a diameter of 4 mm and a length of 20 mm at the indirect head of the model ring. The simulation measures at the model joint are shown in Figures 2 and 3.

Figure 2. Diagram of pipe segment cutting.

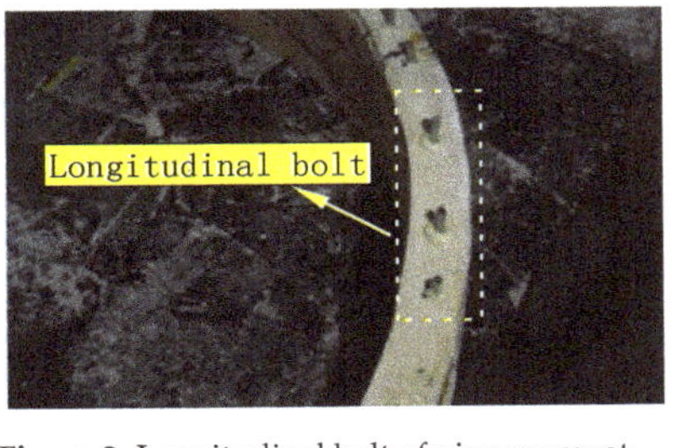

Figure 3. Longitudinal bolt of pipe segment.

The segment similarity model is created by vertically superimposing the segment in the form of one full ring and two half rings (1/2 + 1 + 1/2). The model's overall height is 20 cm (the whole ring is 10 cm high, the half ring is 5 cm), its outer diameter is 0.38 m, and its inner diameter is 0.35 m, as illustrated in Figure 4.

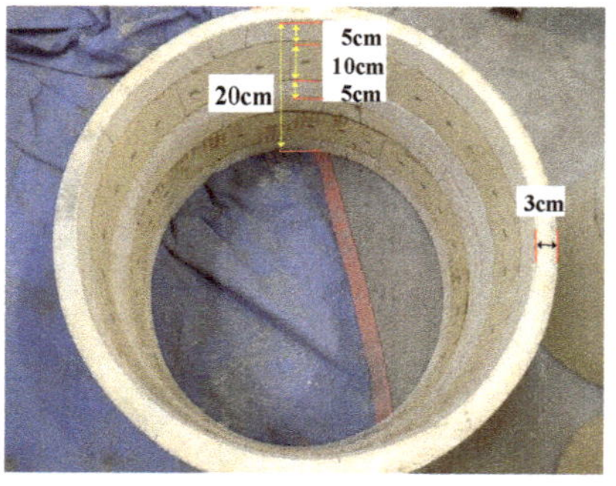

Figure 4. Complete segment model.

2.3. Test Scheme

For the underwater shield tunnel, set the load mode [35] as shown in Figure 5. In the figure, R is the radius of the tunnel. P is the vertical earth pressure. q is the horizontal earth pressure. P_w is the water pressure and P_K is the ground resistance. The water pressure is simulated by hoop force. The concentrated stress force is used to approximately simulate the earth pressure. The concentrated pull force perpendicular to the vertical earth pressure is used to introduce the formation resistance.

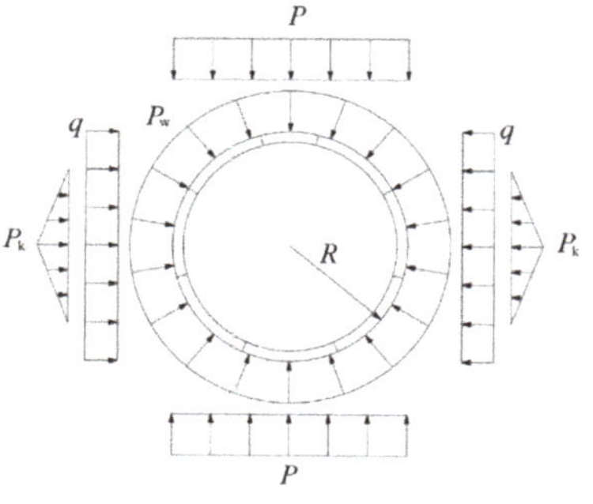

Figure 5. Underwater shield tunnel load diagram.

This experiment examines how shield pipe liners deform when subjected to the combined effects of water and ground pressure. Decide the shield tunnel–stratum similar model test system to load the formation pressure. Select the uniform hydraulic pressure loading device to load water pressure. The shield tunnel–stratum similar model test system as shown in Figure 6. Hydraulic pressure loading device as shown in Figure 7.

According to the real project and the notion of model test resemblance, the model's water pressure was ultimately decided upon as 322 N, and the formation side pressure coefficient as 0.5. The model test will apply formation pressure in three directions, namely the vertical X direction corresponding to the shield tunnel, the lateral Y direction corresponding to the tunnel, and the radial Z direction corresponding to the tunnel, by the shield tunnel–stratum similar model test system. The loading grade is 0.4 MPa per stage, and the vertical formation pressure of the tunnel is loaded incrementally from 0.4 MPa to 6.0 MPa. The loading grade is 0.2 MPa per stage, and the lateral formation pressure of the tunnel is gradually loaded from 0.2 MPa to 3.0 MPa. The radial formation pressure of the tunnel is kept constant at 16.0 MPa to prevent any radial deformation between the pipe segment and the formation. After each stage of the test's formation pressure loading was finished, a steady pressure condition was maintained for five minutes in order to record the segment's deformation data. The loading circumstances are presented in Table 4.

Figure 6. The shield tunnel–stratum similar model test system.

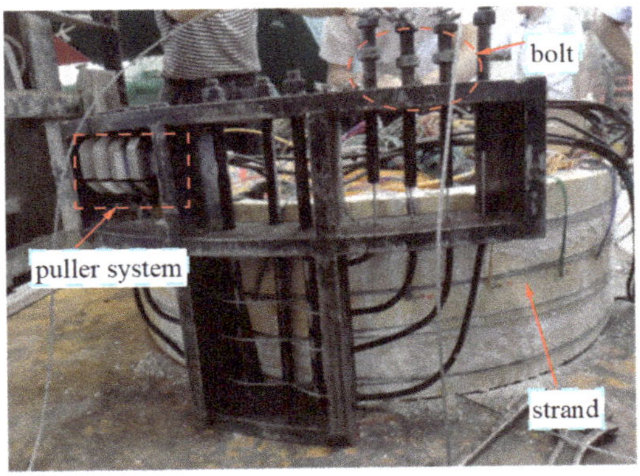

Figure 7. Hydraulic pressure loading device.

Table 4. Test conditions.

Test Water Pressure	Test Surrounding Rock Pressure (Lateral Pressure Coefficient 0.5)		
322 N	Direction I/Vertical (0.4 MPa/level)	Direction II/Lateral (0.2 MPa/level)	Direction III/Radial
	0.4–8.0 MPa	0.2–3.0 MPa	16.0 MPa

Because the similar soil around the segment is not entirely dense during the model loading test, a hydraulic loading device must be used to force the similar soil to act on the surface of the segment once it is fully dense. The hydraulic loading device exhibits a value of 3.6 MPa when the segment is subjected to a complete load, according to data from the entire process recorded by the model test. As a result, the hydraulic device's display of 3.6 MPa in this experiment's model test corresponds to a real stress on the segment of 0 MPa.

2.4. Measuring Scheme

The core part of the fiber grating (FBG) sensing element is the grating, which determines the sensing performance of the sensing element. The central wavelength, grating length, and effective reflectivity are three of the grating's primary characteristic factors that could accurately reflect the sensing performance. Figure 8 and Table 5 display the characteristic parameters and structure diagram of the MMF-G652 fiber grating used in this experiment.

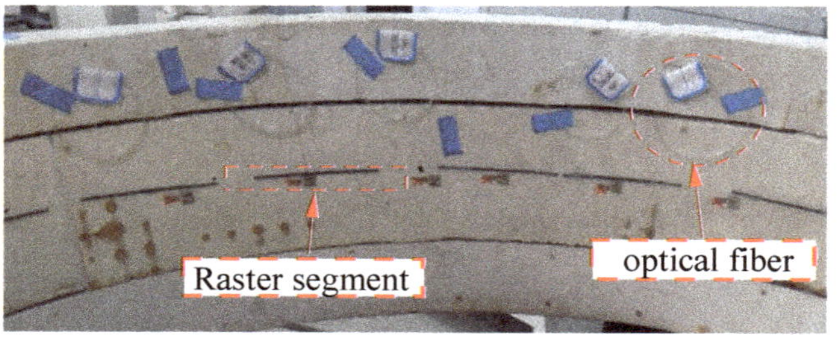

Figure 8. Local layout of the sensing element.

Table 5. Main parameters of fiber grating.

Name	Center Wavelength	Grating Length	Reflection Bandwidth	Array Wavelength Interval	Resolution Ratio	Effective Reflectance
parameter	1510~1560 nm	10 mm	0.2~0.3 nm	5 nm	0.2 pm	95%

In this experiment, the information acquisition form of a surface-mounted FBG sensing element combined with a wavelength demodulator was adopted. The distributed sensor network was formed in the form of circumferential distribution along the inner surface of the segment, and the overall deformation information of the segment structure was accurately and automatically perceived in real time. To fully monitor the tunnel segment for more information, considering the relationship between the length of the optical fiber substrate and the size of the segment, we consider the 18° uniform layout of FBG sensing elements. According to the needs of general engineering monitoring and previous research [34], under the requirements of accuracy and cost, the traditional four-point supplementary layout method is selected. According to the difference between the above two layouts, the third layout is selected as a 36° uniform layout. Therefore, this experiment will use three kinds of distribution methods, and investigate the influence of different distribution methods for surface-mounted FBG sensing elements in the local monitoring of tunnels. The first is an 18° uniform layout mode, in which the FBG sensing element is uniformly arranged in a ring direction on the inner surface of the tube segment, the element's length is chosen to be 8 cm, the ring direction is arranged in a total of 20 elements, and the angle spacing of the element is 18°. The second is a 36° uniform layout mode, in which the component lengths are placed ten times in a ring pattern with a 36° angle between them. The third option is the traditional four-point supplementary layout, in which one measuring point is placed in the vault, one in the bottom of the arch, one in each of the left and right arches, and two more are placed between the four points at angles of 18 degrees and 36 degrees, respectively. In order to eliminate the influence of temperature effect on fiber grating, a temperature compensation fiber is placed near the remaining idle segments. The layout of FBG sensing elements is shown in Figure 9.

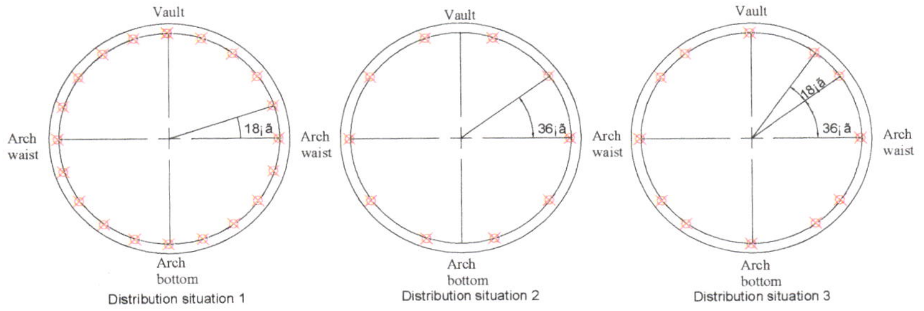

Figure 9. Layout of FBG sensing elements.

3. Tunnel Performance Analysis from Monitoring Results

3.1. Analysis of Segment Strain

Before the loading of surrounding rock pressure, the strain data of the model under no load conditions were measured by the FBG sensing element. Then, the strain data of the model after loading water pressure were measured. Finally, the segment model was loaded according to the test scheme, and the strain information at each position on the inner surface of the model was measured.

The segment's strain information was gathered under varied load and layout situations by the test plan. Figures 10 and 11 illustrate the segment strain following water pressure loading and segment strain during ground loading for Layout 1 (18° uniform layout).

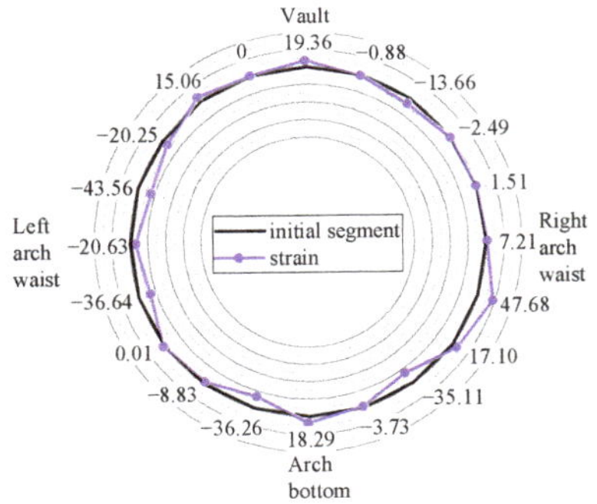

Figure 10. Layout 1: strain distribution of segment after hydraulic loading.

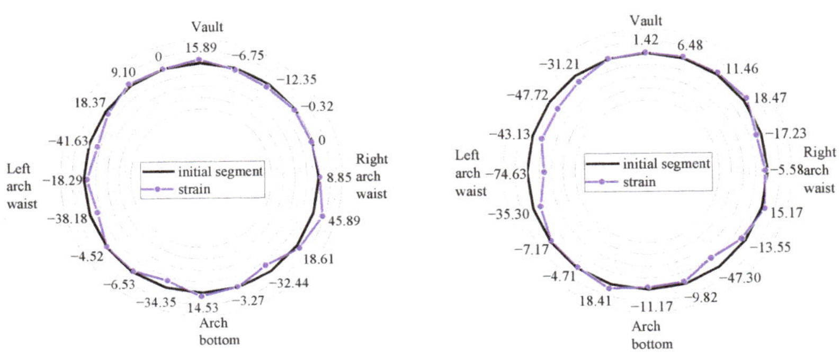

(a) 0.4 MPa

(b) 3.6 MPa

Figure 11. *Cont.*

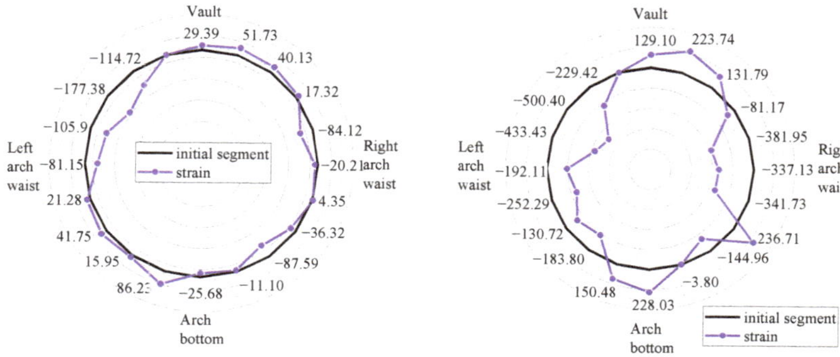

(**c**) 4 MPa (**d**) 4.8 MPa

Figure 11. Layout 1: distribution of ground load loading segment strain.

As seen in Figure 10, when subjected to a 322 N circumferential uniform load, the segment's inner surface experienced a maximum tensile strain of 43.56 $\mu\varepsilon$ and a maximum compressive strain of 47.68 $\mu\varepsilon$. As seen in Figure 11a,b, under the condition of circumferential, uniform water pressure load, the tensile and compressive strains of the segment structure gradually increased again as the formation pressure increased from 0.4 MPa to 3.6 MPa. In general, the strain variation of the segment model is small in the early stage of stratum load loading. Under uniform water pressure, compressive strain occurs on the inner surface of the segment model. The circumferential compressive strain inside the model structure increases. On the whole, the segment has a tendency to shrink inward. The presence of water pressure lessens the strain on the segment lining structure, and its distribution is advantageous to the structure within its permissible range. The strain distribution of the segment exhibits a significant increase when the test loading device is loaded with a 3.6 MPa formation load. The strain on both sides of the segment arch is shifted inward, and the offset gradually increases.

According to the segment diagram of strain distribution, the strain of the segment structure increases continuously during the formation pressure loading process, with compressive strain occurring in the section between the left and right arches and tensile strain occurring between the arch and the arch bottom. The left and right arches protrude outward, while the arch and arch bottom tend to converge inward, in the genuine structural deformation of the segment. The shield tunnel segment's convergence and distortion are elliptical.

In the instance of Layout 2 (36° uniform layout), Figures 12 and 13 illustrate segment strain following water pressure loading and segment strain during ground loading.

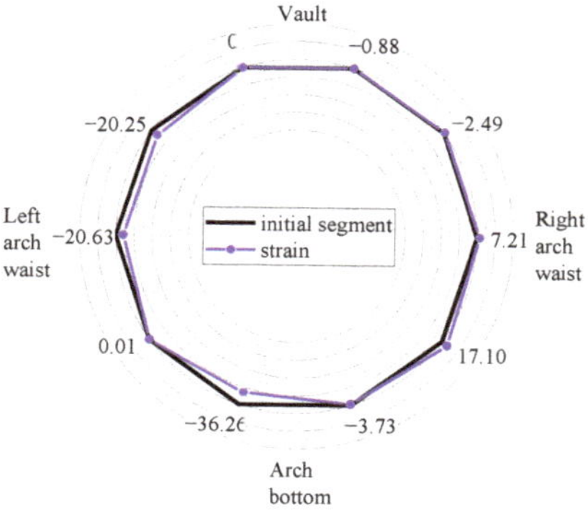

Figure 12. Layout 2: strain distribution of segment after hydraulic loading.

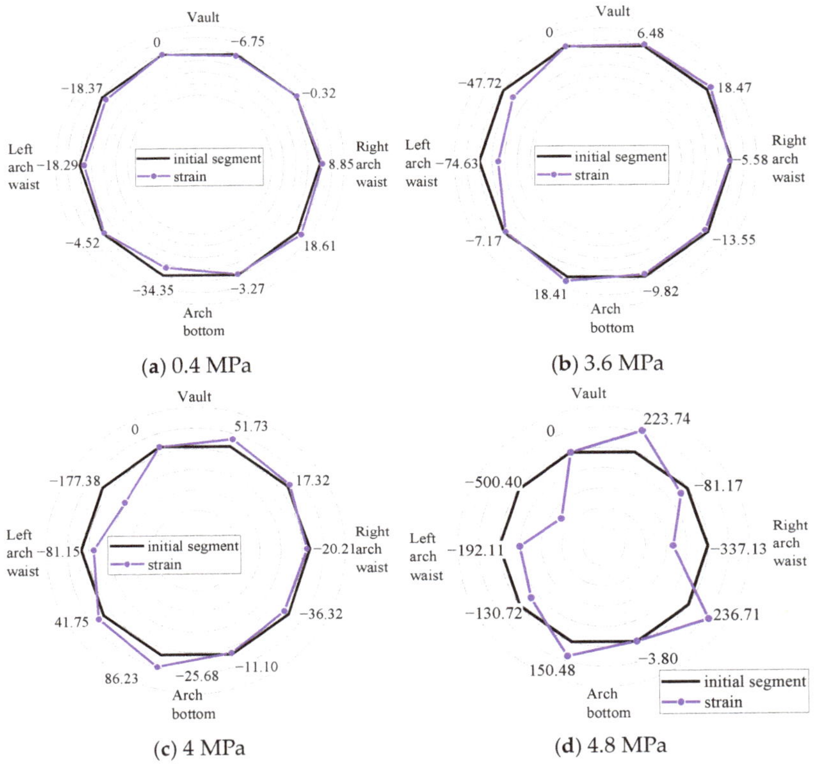

Figure 13. Layout 2: distribution of ground load loading segment strain.

According to Figure 12, when subjected to a 322 N circumferential uniform load, the segment's inner surface experienced a maximum tensile strain of 19.36 με and a maximum compressive strain of 35.11 με. The tensile and compressive strain are not axisymmetric, contrary to Layout 1's finding. The tunnel's general deformation law is the same as Layout 1. It is worth noting that the maximum tensile stress monitored by Layout 1 and Layout 2 is different. In particular, with underwater pressure load and no formation load, the maximum tensile stress of Layouts 1 and 2 is 47.68 MPa and 17.10 MPa, respectively. Layout 1 and Layout 2 have essentially the same maximum compressive stress. In general, the segment lining deforms symmetrically. The top and bottom of the arches contract inward, while the arches on both sides expand outward. Eventually, elliptical distortion can be seen in the segment lining.

When the formation load is 3.6 MPa and 4.8 MPa, the comparison between Layout 1 and Layout 2 reveals that the maximum tensile and compressive stress point of Layout 2 is different from that of Layout 1. The peak value of tensile and compressive stress is not accurately detected, indicating that Layout 1 could more accurately detect changes in tunnel strain.

In the instance of Layout 3 (the traditional four-point supplemental layout), segment strain following water pressure loading and segment strain during ground loading are depicted in Figures 14 and 15.

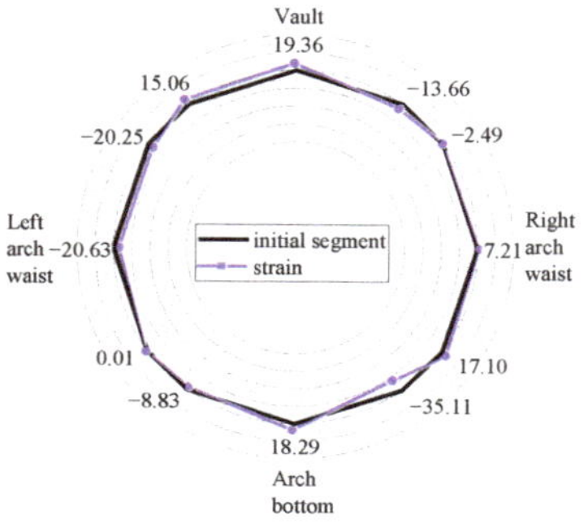

Figure 14. Layout 3: strain distribution of segment after hydraulic loading.

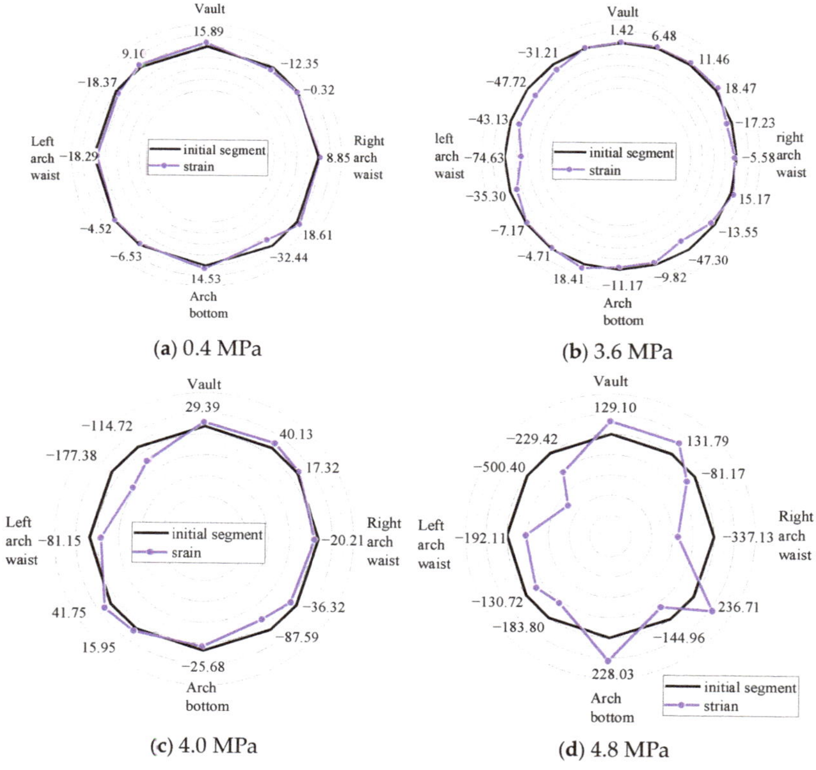

Figure 15. Layout 3: distribution of ground load loading segment strain.

As shown in Figures 14 and 15, under the action of 322 N circumferential uniform load, the maximum tensile strain on the inner surface of the segment model is 19.36 με, and the maximum compressive strain on the surface is 35.11 με. Different from the results of Layout 1, the tensile and compressive strains are not axially symmetrical. As shown in Figure 15 a, d, the maximum compressive strain is 32.44 με, which lowers by 2.67 με when the hydraulic device value increases from 0 to 0.4 MPa. The maximal tensile strain is 15.89 με, which decreases by 3.47 με. The maximum tensile and compressive stress in Layout 3 is the same as that in Layout 1 when the ground load is between 3.6 and 4.8 MPa. The arch top and arch bottom contract inward and the arch waist expands outward on both sides during the deformation of segment lining, which often takes the form of a symmetrical elliptical deformation. Due to the different monitoring arrangements, the measured deformation positions of the tunnel are different, and the deformation of each point of the tunnel is also different. This also leads to differences in measurement data under different monitoring arrangements for the same loading mode. However, the tunnel deformation law of the overall monitoring is the same.

The conclusion of Layout 3 is the same as that of Layout 1 under the reduction in FBG sensing points, and the peak value of tensile and compressive stress is the same under various loads. The comparison of the two layouts demonstrates that Layout 3 could also satisfy the requirements for stress monitoring under the reduction in the number of monitoring points. Thus, Layout 3 is preferred for monitoring tunnel stress parameters.

The maximum tension and compression strains under each load in Layout 1 were utilized as the primary reference indices to construct the segment load–strain curve, appropriately showcasing the structure's deformation state. As demonstrated in Figure 16, the maximum compressive strain of the segment lining structure steadily rises, from −41.63 με to −74.63 με, a 73% increase, when the ground load is in the range of 0–3.6 MPa. The maximum tensile strain initially rises and then gradually falls to around 50 με. The segment structure's maximum compressive and tensile strains rose sharply between 3.6 MPa and 4.8 MPa, with the compressive strain rising by 481% from 74.63 με to 433.43 με. Following the formation load of 4.0 MPa, the segment structure deforms rapidly. The portion has fissures, which eventually progress gradually. Finally, the structure sustains extensive damage from numerous secondary cracks that grow and penetrate.

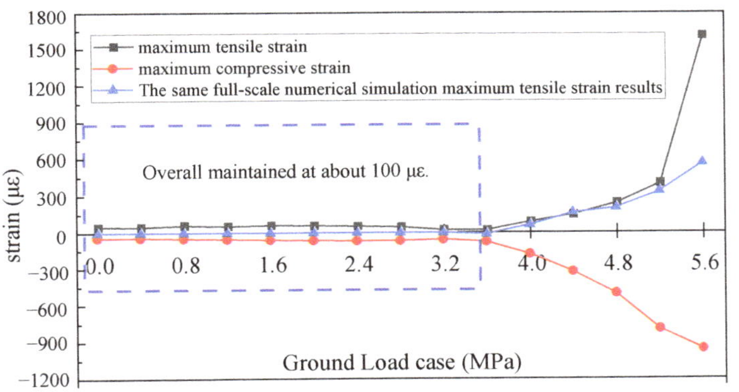

Figure 16. The segment's maximum tensile and compressive strain curve.

Figure 16 shows a comparison between the numerical simulation results and the experimental data of the same full-scale tunnel [36]. The scale test of 0–3.6 MPa is in the stage of soil relaxation. Therefore, compared with the situation after 3.6 MPa, the results of the scaled segment are similar to those of the numerical model between the crack initiation and expansion at 5.2 MPa. After 5.2 MPa, because the numerical model adopts the Moore algorithm, the strain results after the crack attempt cannot be considered. Therefore, the difference between the data is relatively large. This also proves the feasibility of the test results.

Combined with the deformation information of the segment structure provided above, the following conclusions could be derived:

(1) The segment structure benefits from circumferential uniform water pressure and a water pressure of 322 N could effectively limit the structural deformation of the segment under a vertical formation load of 0.4 MPa;

(2) When the segment structure is subjected to formation pressure, the tensile strain of the arch and arch bottom gradually increases, causing the arch and arch bottom of the

structure to shrink inward and deform; the compressive strain of the arch gradually increases, causing both sides of the arches to bulge out and deform;

(3) For all three layouts, the segments' derived strain laws are essentially the same. Layout 3 (traditional four-point supplementary layout) performs better when the strain of the segments is monitored by the FBG sensing element. Layout 3 could effectively gather data on structural strain when the number of layout points is minimized.

3.2. Analysis of Segment Curvature

The initial curvature, as determined by the test methodology, was 0.0253 m^{-1} before the start of the formation pressure loading. This curvature was determined by using the relationship between the radius and curvature. The segment was then loaded by the test plan, and the wavelength drift following the loading of the earth pressure was determined by the surface-mounted FBG sensing element. The relationship between the wavelength drift and the curvature was used to calculate the curvature information for each place on the inner surface of the model.

Figure 17 depicts the variance in segment curvature during formation loading for Layout 1 (18% uniform layout).

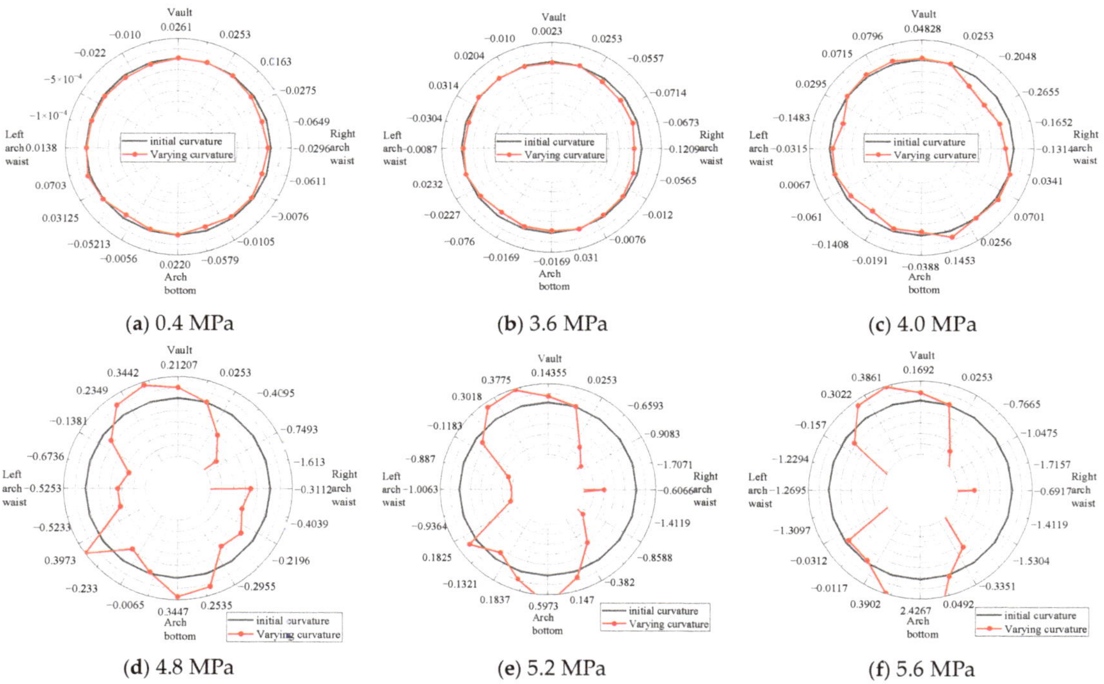

Figure 17. Layout 1: segment curvature distribution following ground loading.

As seen in Figure 17a,b, at this point, formation pressure gradually affected the segment. The inner surface of the segment exhibited primarily negative curvature growth changes, indicating that more pressure strain occurred inside the model's structure, and the segment as a whole displayed a trend of inward contraction and deformation. In the early stages of formation loading, the segment curvature does not vary much. The curvature of the segment dramatically alters when the test loading device is loaded with a 3.6 MPa formation load, as shown in Figure 17c–f. The segment's highest curvature, with 4.8 MPa ground load, is 0.3973 m^{-1}, while its smallest curvature is -1.613 m^{-1}. This is because as ground load increases, the segment's curvature on both sides of the arch waist grows negatively, indicating an outward displacement, while the segment's curvature in the vault and arch bottom area grows positively, indicating an inward displacement.

According to the curvature distribution diagram of the segment under Layout 1, in the process of formation pressure loading, positive curvature changes occur in the section of the arch and arch bottom and negative curvature grows in the section of the left and right arch waist. The left and right arch waist protrude outward and the arch and arch bottom converge inward as part of the segment's actual structural deformation.

Figure 18 depicts the variance in segment curvature during formation loading for Layout 2 (18% uniform layout).

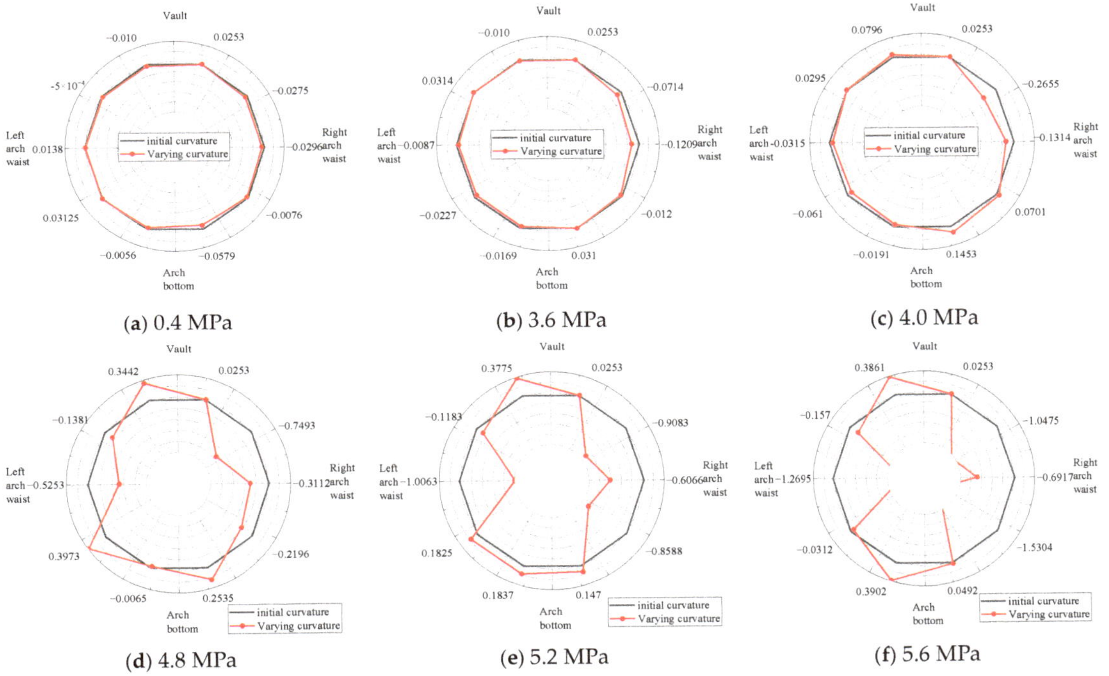

Figure 18. Layout 2: segment curvature distribution following ground loading.

In Figure 18a–f, during the formation pressure loading process, positive curvature changes happen in the section of the arch top and arch bottom while negative curvature increases in the section of the left and right arch waistline. Additionally, the deformation state of the shield tunnel segments exhibits convergence and deformation. Layout 2 shares the same variation in tunnel curvature distribution as Layout 1. By comparing the distribution of the tunnel's maximum and minimum curvature under Layout 1 and Layout 2, it could be seen that the distribution of the maximum and minimum curvature is the same under the formation loading of 4 MPa, while the other loads are different. The maximum curvature of Layout 2 at 5.6 MPa is 0.3902 m^{-1}, while the maximum curvature of Layout 1 is 2.4267 m^{-1}, demonstrating that Layout 2's monitoring of local curvature changes is less precise than that of Layout 1. The findings indicate that Layout 1 is superior at monitoring segment curvature because it is not accurate to examine tunnel deformation using the curvature change data gathered in Layout 2 compared to Layout 1.

Figure 19 depicts the variance in segment curvature during formation loading for Layout 3 (the traditional four-point supplemental layout).

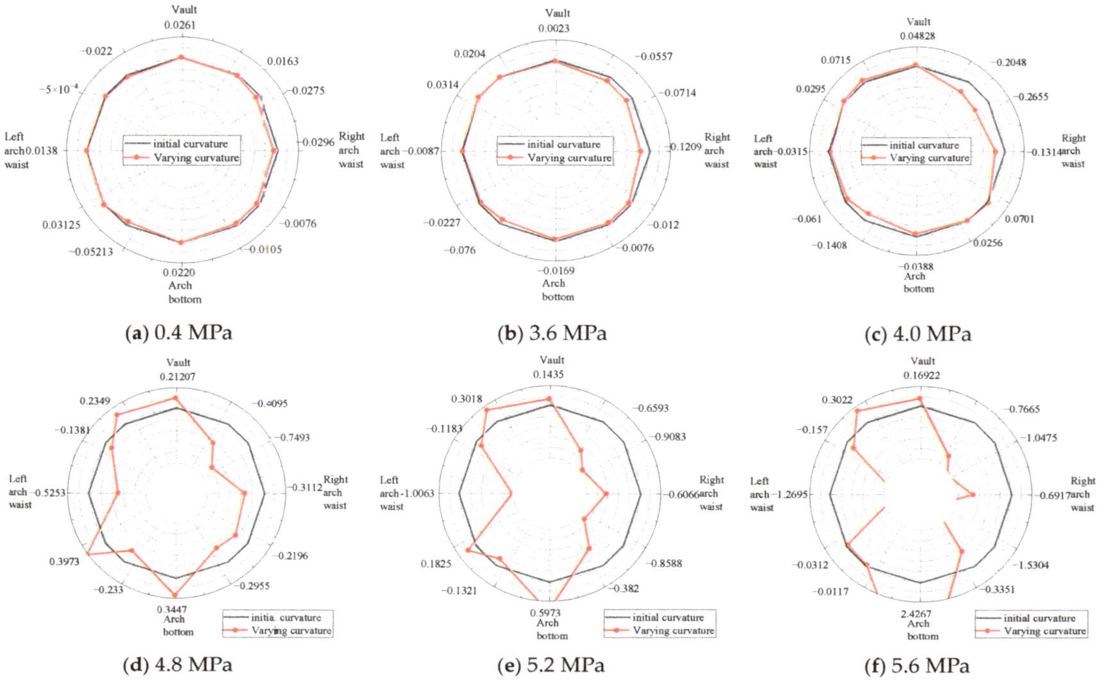

Figure 19. Layout 3: segment curvature distribution following ground loading.

Figure 19 illustrates how formation pressure loading causes positive curvature changes in the arch top and arch bottom sections while increasing negative curvature in the left and right arch waist sections. The tunnel curvature distribution alteration rule is the same as Layout 1. According to Figure 19 a, d, the tunnel's maximum curvature and lowest curvature distribution are compared for Layouts 1 and 3. The distribution of maximum curvature and lowest curvature measurement locations under the two types of layouts is different in the early stages of formation loading, as can be observed, but the difference in peak values is negligible, thus the influence on local curvature monitoring may be disregarded. Figure 19e–f shows that, for ground loads of 5.2 MPa and 5.6 MPa, the maximum curvature distribution is the same as the maximum curvature distribution in Layout 1; however, the extreme curvature position varies for the other ground load.

The deformation state of shield tunnel segments exhibits a convergence deformation under the curvature change in Layout 1 and Layout 3; this can be discovered by contrasting the two layout circumstances of Layout 1 and Layout 3. However, in the case of Layout 3, the position distribution of extreme curvature at the late loading stage is different from that of Layout 1, indicating that Layout 3 is unable to accurately collect the local curvature of tunnel curvature, while Layout 1 is better at monitoring the curvature of segments.

The minimum curvature data varied significantly overall, as shown by the tunnel curvature distribution maps of the three layout modes mentioned above. The maximum curvature under Layout 1 was used as the primary reference index to describe the structure's state of deformation, and the maximum curvature of the segment was shown as a load–curvature change curve, as shown in Figure 20.

Figure 20 illustrates how ground load causes the segment curvature to progressively increase in the shape of a 'ladder'.

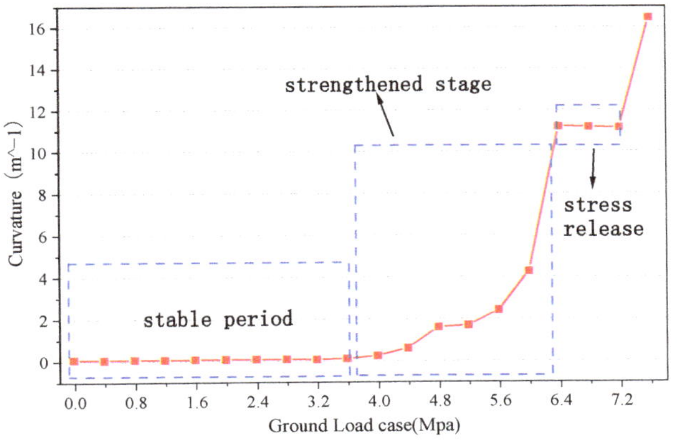

Figure 20. Maximum curvature change in the segment load curve.

The curvature change exhibits a clear gradient increase trend when combined with the deformation process information of the segment previously discussed and shows that the total change trend of the maximum curvature of the segment can be separated into five stages. The segment structure primarily deforms into an elastic form in the first stage (0–3.6 MPa), where the curvature is essentially maintained near 0.1 m^{-1}. The maximum curvature gradually increases from 0.1209 m^{-1} to 2.429 m^{-1} with a slope of 1.2 m^{-1}/MPa in the second stage (3.6–5.6 MPa). It should be noted that, after 4.8 MPa, the curvature began to rise rapidly. This is because the crack propagation occurs in the segment, which corresponds to the stress release stage of the strain curve. The greatest curvature increases significantly from 2.429 m^{-1} to 11.2037 m^{-1} with a slope of 10.97 m^{-1}/MPa in the third stage (5.6–6.4 MPa). Subtle fractures begin to show up inside the structure as the second and third stages of the segment's strengthening progress, and the fissures steadily deepen as the load increases. The fourth stage (6.4–7.2 MPa) sees the development of internal fissures and the release of tension. After 7.2 MPa, the segment structure enters the fifth stage, when cracks continue to grow, widen, and penetrate until the segment is no longer able to support the weight.

The curvature of the arch measuring point, which is the important position of the segment, is used as an example to evaluate the deformation rule of the shield tunnel segment structure because the arch is a frequent deformation notable place in engineering. Figure 21 depicts the load–curvature change curve of the segment measuring point for the arch roof.

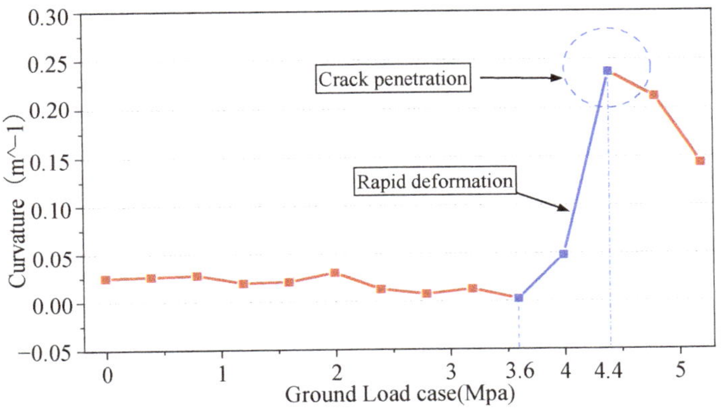

Figure 21. Segment arch's curve of curvature under load.

In the load range of 0–3.6 MPa, Figure 21 shows that the segment measuring point arch's curvature steadily reduces. The fracture was broken through and the structure tension was released when the vertical formation load was 4.4 MPa; the curvature of the arch abruptly decreased (slope 0.3), indicating that the fracture had finally occurred.

The following conclusions could be drawn from the study presented above. Under the combined action of ground load and water pressure, the curvature of the segment lining changes in a 'ladder' shape, which increases with the increase in ground load. The strengthening effect of constant dispersed water pressure on the segment structure's deformation resistance is more noticeable the smaller the ground load and the effect of uniform water pressure is negated when the ground load surpasses 4.0 MPa. An incompletely symmetrical outward spreading deformation occurs in the left spandrel area of the segment lining when uniform water pressure is applied to the formation load below 4.0 MPa. The derived segment curvature change law is the same for all three layout situations. Layout 1 (18° uniform layout mode) performs better under the monitoring segment curvature condition. It is advised to use a 18° uniform layout of FBG sensing elements for information collection when monitoring tunnel sections in challenging terrains.

4. Conclusions

This study thoroughly examines the impact of various layout schemes on the accurate monitoring of local tunnel deformation and the overall deformation law of shield segments under the action of water pressure and formation load; three surfacing FBG sensing element layout schemes were adopted. A 1:20-scale model test was carried out using a rotary hydraulic loading device and the shield tunnel formation complicated the simulation test system. In this study, the deformation characteristics of tunnel segments were studied by combining two kinds of change information on strain and curvature. This experiment only studied the influence of strain and curvature under different FBG layouts for fixed water pressure and variable soil pressure. The following conclusions were obtained.

(1) In the 1:20-scale model test, uniform water pressure causes the entire segment to contract inward into a 'circular' shape, while strata load gradually deforms the segment into an elliptical shape. The structural deformation of the segment lining is asymmetrical for ground loads between 0.4 and 3.6 MPa. The segment lining's left spandrel has a propensity to enlarge outward, whilst the right spandrel has a propensity to contract inward. The deformation of the segment lining restores symmetry after the ground load period of 3.6 MPa.

(2) In the experiment, the curvature change shows a 'ladder' transformation as a whole, which can be divided into three major stages. In the 0–3.6 MPa stage, the segment is in the elastic stage, and the curvature is basically guaranteed to be 0.1 m^{-1}/MPa. In the 3.6–6.4 MPa stage, the segment is in the strengthening stage and the curvature is greatly improved, up to 10.97 m^{-1}/MPa. After 6 MPa, the stress is released and the curvature is further improved after stabilization.

(3) The appearance of cracks after 4.4 MPa has an important influence on the curvature change. The percentage of curvature change in the 0–4.4 MPa stage is 188%, but the percentage of curvature change in the 4.4–8.0 MPa stage is as high as 2900%.

(4) Before cracks appear in the scale model, the strain and curvature of the tunnel segment increase steadily in a small range. When cracks appear, the values of strain and curvature fluctuate obviously. The combination of the strain and curvature of the shield tunnel segment structure can well reflect the deformation information of the structure.

(5) According to research on the effects of three surface-mount FBG sensing elements, the traditional four-point supplementary layout could more accurately perceive tunnel strain changes, and the 18° uniform layout could more accurately perceive local tunnel curvature changes. It is recommended that the FBG sensing elements with a uniform layout of 18° be used to collect data for monitoring tunnel sections in harsh terrain and the FBG sensing elements with a four-point additional configuration be used to collect data for monitoring other sections.

In this study, a 1:20 scale model test was carried out to investigate the deformation perception of the shield tunnel structure under three different layouts of surface-mounted FBG sensors. Due to the influence of scale effect, the layout method will have errors in practical engineering, which need to be verified in the next step. The above research focuses on the section form of the shield tunnel, and the horseshoe tunnel needs to be further studied. The wall thickness of the tunnel designed in this experiment is a fixed value, and it is not used as a parameter to study the critical pressure. The tunnel wall thickness should be further studied as a variable parameter. The focus of the test is to study the influence of hydrostatic pressure and changing earth pressure on tunnel deformation. The influence of live load on tunnel deformation needs to be further studied.

Author Contributions: Conceptualization, C.H., R.X. and P.X.; Methodology, R.X., D.Z. and Y.L. (Yu Liu); Software, Y.L. (Yufu Li). and Z.Z.; Validation, R.X.; Formal Analysis, Y.L. (Yufu Li); Investigation, Y.L. (Yufu Li); Resources, C.H. and R.X.; Data Curation, Y.L. (Yufu Li) and Y.L. (Yujun Li); Writing—Original Draft Preparation, R.X., Y.L. (Yufu Li) and D.Z.; Writing—Review & Editing, Y.L. (Yufu Li), D.Z. and R.X.; Visualization, Y.L. (Yufu Li) and Y.L. (Yujun Li); Supervision, R.X., C.H. and K.F.; Project Administration, C.H., R.X. and P.X.; Funding Acquisition, R.X. and Y.L. (Yufu Li). All authors have read and agreed to the published version of the manuscript.

Funding: This research was funded by the General Fund of Chongqing Natural Science Foundation (No. CSTB2023NSCQ-MSX0960), the State Key Laboratory of Mountain Bridge and Tunnel Engineering Fund Project (No. SKLBT-2109), the Major Science and Technology Special Project of Xinjiang Uygur Autonomous Region (No. 2020A03003-4), the Postdoctoral Special Funding Project of Chongqing (No. 01.02.9901212074), and the Research and Innovation Program for Graduate Students in Chongqing (No. CYS240457).

Data Availability Statement: The data presented in this study are available on request from the corresponding author. The data are not publicly available due to the restriction of privacy.

Conflicts of Interest: The authors declare no competing interests.

References

1. Hong, C.Y.; Zhang, Y.F.; Li, G.W.; Zhang, M.X.; Liu, Z.X. Recent progress of using Brillouin distributed fiber optic sensors for geotechnical health monitoring. *Sens. Actuators A Phys.* **2017**, *258*, 131–145. [CrossRef]
2. Jali, M.H.; Rahim, H.R.A.; Johari, M.A.M.; Baharom, M.F.; Ahmad, A.; Yusof, H.H.M.; Harun, S.W. Optical Microfiber Sensor: A Review. *J. Phys. Conf. Ser.* **2021**, *2075*, 012021. [CrossRef]
3. Rao, Y.-J. Recent Progress in Ultra-Long Distributed Fiber-Optic Sensing. *Acta Phys. Sin.* **2017**, *66*, 139–157.
4. Westbrook, P.S.; Kremp, T.; Feder, K.S.; Ko, W.; Monberg, E.M.; Wu, H.; Simoff, D.A.; Taunay, T.F.; Ortiz, R.M. Continuous multicore optical fiber grating arrays for distributed sensing applications. *J. Light. Technol.* **2017**, *35*, 1248–1252. [CrossRef]
5. Yüksel, K.; Kinet, D.; Moeyaert, V. Railway Monitoring System Using Optical Fiber Grating Accelerometers. *Smart Mater. Struct.* **2018**, *27*, 105033. [CrossRef]
6. Sun, L.; Li, C.; Zhang, C.; Su, Z.; Chen, C. Early monitoring of rebar corrosion evolution based on FBG sensor. *Int. J. Struct. Stab. Dyn.* **2018**, *18*, 1840001. [CrossRef]
7. Li, C.; Sun, L.; Xu, Z.; Wu, X.; Liang, T.; Shi, W. Experimental investigation and error analysis of high precision FBG displacement sensor for structure health monitoring. *Int. J. Struct. Stab. Dyn.* **2020**, *20*, 2040011. [CrossRef]
8. Braunfelds, J.; Senkans, U.; Skels, P.; Janeliukstis, R.; Salgals, T.; Redka, D.; Lyashuk, I.; Porins, J.; Spolitis, S.; Haritonovs, V.; et al. FBG-based sensing for structure health monitoring of road infrastructure. *J. Sens.* **2021**, *2021*, 8850368. [CrossRef]
9. Wang, T.; Tang, Y.; Yang, H.; Xu, X.; Liu, W.; Li, X. Convergence deformation monitoring of a shield tunnel based on flexible long-gauge FBG sensors. *Mech. Adv. Mater. Struct.* **2022**, *29*, 2827–2835. [CrossRef]
10. Liang, M.; Fang, X.; Song, Y.; Li, S.; Chen, N.; Zhang, F. Research on Three-Dimensional Stress Monitoring Method of Surrounding Rock Based on FBG Sensing Technology. *Sensors* **2022**, *22*, 624. [CrossRef]
11. Liang, M.F.; Chen, N.; Fang, X.Q.; Wu, G. Strain Transferring Mechanism Analysis of the Surface-Bonded FBG Sensor. *Appl. Opt.* **2018**, *57*, 5837–5843. [CrossRef] [PubMed]
12. Sigurdardottir, D.H.; Stearns, J.; Glisic, B. Error in the Determination of the Deformed Shape of Prismatic Beams Using the Double Integration of Curvature. *Smart Mater. Struct.* **2017**, *26*, 075002. [CrossRef]
13. Nawrot, U.; Geernaert, T.; De Pauw, B.; Anastasopoulos, D.; Reynders, E.; De Roeck, G.; Berghmans, F. Development of a Mechanical Strain Amplifying Transducer with Bragg Grating Sensor for Low-Amplitude Strain Sensing. *Smart Mater. Struct.* **2017**, *26*, 075006. [CrossRef]

14. Liu, Y.; Feng, Y.; Wen, J.; Huang, L.; Dong, J. Integrated fiber-optic sensor based on the inscription of FBG in seven-core fiber for curvature and temperature measurements. *Opt. Fiber Technol.* **2023**, *75*, 03197. [CrossRef]
15. Sun, X.; Wang, Q.; Li, Z.; Chu, J. Abnormal diagnosis of fiber Bragg grating strain gauge for structural health monitoring system. *Opt. Precis. Eng.* **2021**, *29*, 2581–2589. [CrossRef]
16. Zhou, M.; Wang, C.; Huang, H.; Cheng, W.; Shao, H.; Zhang, D. Safety state evaluation of operational shield tunnel structure based on machine vision and performance analysis. *J. Appl. Basic Eng. Sci.* **2023**, *31*, 1461–1476.
17. Chen, X.; Wang, F.; Wu, M.; Wang, Y. All-round monitoring method of tunnel section deformation based on inclination inversion. *J. Cent. South Univ. (Nat. Sci. Ed.)* **2024**, *55*, 151–161.
18. Zhou, T.; Zou, J.; Guo, J. Research on tunnel deformation monitoring and intelligent early warning method. *Bull. Surv. Mapp.* **2022**, 91–94.
19. Sui, Y.; Cheng, X.; Li, G.; Pu, L.; Li, C.; Liao, P. Inversion analysis of deformation stress of shield tunnel segments based on distributed optical fiber monitoring. *Eng. Mech.* **2022**, *39*, 158–163.
20. Zheng, T. Method for shield tunnel cross-section curve reconstruction based on FBG sensors and neural network. *Acad. J. Comput. Inf. Sci.* **2022**, *5*, 15–21.
21. Letsios, V.; Faraslis, I.; Stathakis, D. Monitoring Building Activity by Persistent Scatterer Interferometry. *Remote Sens.* **2023**, *15*, 950. [CrossRef]
22. Roccheggiani, M.; Piacentini, D.; Tirincanti, E.; Perissin, D.; Menichetti, M. Detection and Monitoring of Tunneling Induced Ground Movements Using Sentinel-1 SAR Interferometry. *Remote Sens.* **2019**, *11*, 639. [CrossRef]
23. Xu, L.; Gao, G.; Peng, C.; Wang, T.; Yang, J.; Xie, J. Train structural health monitoring method based on fiber optic sensing and piezoelectric sensing. *Chin. J. Railw. Sci. Eng.* **2023**, *20*, 2763–2772.
24. Ghaderpour, E.; Mazzanti, P.; Bozzano, F.; Mugnozza, G.S. Ground deformation monitoring via PS-InSAR time series: An industrial zone in Sacco River Valley, central Italy. *Remote Sens. Appl. Soc. Environ.* **2024**, *34*, 101191. [CrossRef]
25. Tang, Z.C.; He, C.; Lin, G. Study on the mechanical behavior of segment structure of shield tunnel of metro with model test. *Yantu Gongcheng Xuebao (Chin. J. Geotech. Eng.)* **2005**, *27*, 85–89.
26. He, C.; Feng, K.; Yang, X. Model test on segmental lining of Nanjing Yangtze River tunnel with super-large cross-section. *Chin. J. Rock Mech. Eng.* **2007**, *26*, 2260–2269.
27. Kang, C.; Mei, G.X.; Liang, R.Z.; Wu, W.B.; Fang, Y.X.; Ke, Z. B Analysis of the longitudinal deformation of existing shield tunnel induced by temporary surface surcharge. *Rock Soil Mech* **2018**, *39*, 4605–4616.
28. Ding, Z.; Zhang, X.; Yin, X.S.; Jiang, J.Q. Analysis of the Influence of Soft Soil Grouting on the Metro Tunnel Based on Field Measurement. *Eng. Comput.* **2019**, *36*, 1522–1541. [CrossRef]
29. Zhang, Z.G.; Huang, M.S.; Zhang, C.P.; Jiang, K.M.; Bai, Q.M. Analytical Prediction of Tunneling-Induced Ground Movements and Liner Deformation in Saturated Soils Considering Influences of Shield Air Pressure. *Appl. Math. Model.* **2020**, *78*, 749–772. [CrossRef]
30. Wang, S.M.; Wang, X.M.; Chen, B.; Fu, Y.B.; Jian, Y.Q.; Lu, X.X. Critical State Analysis of Instability of Shield Tunnel Segment Lining. *Tunn. Undergr. Space Technol.* **2020**, *96*, 103180. [CrossRef]
31. Tan, Z.S.; Li, Z.L.; Tang, W.; Chen, X.Y.; Duan, J.M. Research on Stress Characteristics of Segment Structure during the Construction of the Large-Diameter Shield Tunnel and Cross-Passage. *Symmetry* **2020**, *12*, 1246. [CrossRef]
32. Civera, M.; Dalmasso, M.; Chiaia, B. Assessing the Seismic Performance of Underground Infrastructures to Near-Field Earthquakes. *Int. J. Civ. Infrastruct.* **2024**, *2024*, 32–42. [CrossRef]
33. He, C.; Zhang, J.G.; Yang, Z. Model test study on the mechanical characteristics of segment lining for the Wuhan Yangtze River tunnel. *China Civ. Eng. J.* **2008**, *41*, 85–90.
34. Zhang, J.G.; He, C.; Yang, Z. Study on mechanical distribution characteristics of segment lining structure affected by width in Wuhan Yangtze River Tunnel. *Chin. J. Rock Mech. Eng.* **2007**, *S2*, 3763–3769.
35. He, C.; Feng, K.; Su, Z. Development and application of loading test system for prototype structure of large-section underwater shield tunnel. *Chin. J. Rock Mech. Eng.* **2011**, *30*, 254–266.
36. Yao, Z. *Research on Deformation State Perception of Underwater Shield Tunnel Structure Based on Fiber Grating Sensing*; Chongqing Jiaotong University: Chongqing, China, 2022.

buildings

MDPI

Article

Study on Deformation Characteristics of the Segment in the Underwater Shield Tunnel with Varying Earth Pressure

Rongjun Xing [1,2], Yujun Li [1,2], Chuan He [3], Daiqiang Zhu [1,2], Yufu Li [1,2], Chaofan Yao [3], Yu Liu [1,2], Pai Xu [1,2,*] and Zhongchao Zhao [1,2]

1 State Key Laboratory of Mountain Bridge and Tunnel Engineering, Chongqing Jiaotong University, Chongqing 400074, China; xingrong-12@163.com (R.X.); m15730446752@163.com (Y.L.); zhu_daiqiang@163.com (D.Z.); 18284886470@163.com (Y.L.); 15215150872@163.com (Y.L.); zc-zhao@mails.cqjtu.edu.cn (Z.Z.)

2 School of Civil Engineering, Chongqing Jiaotong University, Chongqing 400074, China

3 Key Laboratory of Transportation Tunnel Engineering, Ministry of Education, Southwest Jiaotong University, Chengdu 610031, China; chuanhe21@163.com (C.H.); yaochaofan@hotmail.com (C.Y.)

* Correspondence: xu_pai@126.com

Abstract: The segment in an underwater shield tunnel is influenced by the change of earth pressure and water pressure. Therefore, the law of segment deformation should be mastered for the safe operation of the tunnel. To obtain the law of segment deformation under varying earth pressure, loading conditions of constant water pressure and without water pressure were considered. In this study, the numerical simulation and scale model experiment were carried to analyze the strain, curvature, and displacement of the segment. The results show that the strain amplitude of segments is reduced by the water pressure under a range of earth pressure. When the earth pressure ranges from 0 MPa to 2.4 MPa and the water pressure 0.33 MPa, the displacement of the vault and arch waist exhibit a decreased rate of 7.916 mm/MPa and an increased rate of 5.416 mm/MPa, respectively. Under the combined effects of constant water pressure and varying earth pressure, the curvature of the segment tends to stabilize after a rapid change with a maximum of 0.004 m^{-1}.

Keywords: underwater shield tunnel; numerical simulation; scale model experiment; shield segment deformation; variable earth pressure

1. Introduction

In recent years, shield tunneling has been widely used in the construction of underwater tunnels due to its cost-effectiveness and small disturbance to the surrounding soil [1]. However, due to the complexity of the hydrological conditions and geological environment of underwater shield tunnels, there are inevitable phenomena such as crack propagation, deformation accumulation, and resistance attenuation in the tunnel structure, which affect the health status of the shield tunnel segment during the operation stage. It is prone to occur in accidents in tunnel engineering operations [2]. For example, the Kaohsiung Harbor Crossing Submarine Tunnel in Taiwan experienced problems such as structural cracks and water leakage after 20 years of operation. Therefore, it is of great significance to investigate the deformation law of the shield segment for the safety of underwater shield tunnels.

At present, significant achievements have been made in the study of structural deformation of shield tunnels. Feng et al. [3] conducted large-scale model tests based on the Guangzhou Shiziyang Tunnel project to explore the impact of surrounding rock degradation on the mechanical performance of underwater shield tunnel segment structures. Feng et al. [4] conducted experimental and numerical studies on the bending performance of large underwater shield tunnel segment joints. The results showed that the nodes exhibited linear opening and nonlinear deformation after concrete cracking, and the contact surface of the nodes during the opening process was curved. Wang et al. [5] studied the effects of segment joints, segment sizes, and geological conditions on the buckling of shield tunnel lining under hydrostatic pressure through analytical and numerical analysis and concluded that ground support significantly improves the buckling strength, while earth pressure reduces the ability to resist hydrostatic buckling. Ye et al. [6] conducted model tests on three types of segmented ring test models: straight seam ring, staggered seam ring, and uniform ring. They found that the transverse bending stiffness value of the staggered seam assembly ring was significantly greater than that of the straight seam assembly ring, and this difference gradually decreased with the increase of load. Zhang et al. [7] studied the mechanical properties, deformation, stress distribution, crack development process, and failure mode of mining tunnel lining in aquifers under the combined pressure of water and earth through a 1:30 large-scale model test. Under two test conditions, namely only water pressure and variable water pressure but constant earth pressure, cracks and failure occurred successively at the bottom of the wall and the inverted arch.

Yang et al. [8] analyzed the effects of water level changes, seasonal environmental temperature changes, and time effects on tunnel behavior through a multiple linear regression model. The regression results indicate that temperature is the most important factor affecting the strain of the segment during normal operation, and the annual irreversible deformation of the segment joint is as significant as the irreversible deformation caused by changes in water level and temperature. Arnau and Molins [9] used numerical simulations to consider the three-dimensional response generated by local loads and identified the most important influencing parameters and their effects on the response of the lining structure. They pointed out that the stiffness of the surrounding strata determines the degree of interaction between adjacent rings, while the longitudinal compression of the lining determines the maximum load that produces complete interaction. Cheng et al. [10] investigated the influence of excavation depth ratio and unloading ratio on tunnel displacement behavior through a series of numerical simulations and concluded that the axial force at or near the tunnel sidewall is the most sensitive to the unloading effect caused by excavation activities. Liu et al. [11] designed and implemented a full-scale test on the bearing capacity of staggered assembled shield tunnel structures based on unloading conditions. Under the design load, the progressive failure and ultimate bearing capacity of the staggered segment lining were obtained. Based on considering the elastic-plastic properties of bolts and concrete, Cheng et al. [12] provided analytical solutions for the longitudinal equivalent bending stiffness and shear stiffness of circumferential nodes and proposed two shear load transfer theoretical models for the arrangement of circumferential nodes. Zhang et al. [13] conducted a series of full-scale experimental studies on the failure characteristics of the key segment in super-large cross-section shield tunnels under general loads. The results showed that the longitudinal force acting along the tunnel axis significantly limits the inward convergence of the segment, the opening of circumferential seams, and the displacement of longitudinal and circumferential seams within a certain range. Zhang et al. [14] compared and analyzed the characteristics of structural instability and failure of two types of segment structures under high water pressure through prototype tests of straight seam assembly structure and staggered seam assembly structure segment in the Shiziyang Tunnel. Deng et al. [15] pointed out that concrete structures operating in a water environment for extended periods are in a state of long-term water saturation. The free water present in the concrete pores exerts a certain water pressure on the pore walls under various loads, thereby affecting the static strength of the concrete. Yang et al. [16] studied the effect characteristics of concrete under different water-confining pressures after cyclic loading. Sheng et al. [17] summarized that the concrete structure operating in the water environment exhibits different damage laws and mechanisms under dynamic loading due to the influence of free water. Therefore, it is of great significance to the study of the mechanical characteristics of underwater shield tunnel structures.

From this, it can be seen that the structural deformation and settlement laws of shield tunnels under the influence of a single factor of water pressure or earth pressure are clear. However, under the combined action of water pressure and formation pressure, the stress on the segment structure is indeterminate, and the deformation mechanism is complex. The structural deformation law is not yet clear under the conditions of dual factor cross fusion, especially under the influence of constant water pressure and earth changes pressure. Previous studies often used single physical quantities such as displacement and strain to describe deformation results, which made it difficult to accurately reflect the deformation state of the tunnel segment. Based on this, to make the test conclusion more accurate, the results of the scale test and numerical simulation should be used to verify each other. Firstly, in Section 2.1, the design of numerical simulation parameters is introduced, as well as the setting of measuring points and working conditions. Subsequently, in Section 2.2, the preparation and test scheme for the scale test materials are detailed. In order to obtain the deformation characteristics of underwater shield tunnel segments, in Section 3, the structural strain, curvature, and displacement of the shield tunnel segments under different earth pressures, in both conditions with and without water pressure, were compared and analyzed through numerical simulation and scale model experiments. In Section 4, the deformation characteristics of the shield tunnel segment under the combined action of water pressure and earth pressure are summarized, which provides a reference for the health monitoring, damage prediction, and disaster warning threshold setting of underwater shield tunnels. In addition, this study can provide guiding significance for selecting the appropriate tunnel burial depth in the stratum environment with little difference in water pressure in the planning of the shield tunnel.

2. Numerical Simulation and Scale Model Experiment

2.1. Numerical Simulation of Underwater Shield Tunnels

2.1.1. Model Dimensions and Boundary Conditions

Based on the Jinan Yellow River Shield Tunnel, a shield tunnel segment model was established in ANSYS 16.0. The stratum that the shield tunnel passes through is mainly silty clay, locally with calcareous nodules and lens-shaped fine sand and medium sand strata. The lining rings are 15.2 m in outer diameter, 13.9 m in inner diameter, 0.65 m thick, and 2.0 m wide. C60 reinforced concrete is used.

Taking the longitudinal direction of the tunnel as the Z-axis and the vertical direction as the Y-axis, the Y-axis is rotated 90 degrees clockwise to obtain the X-axis. The boundary dimensions of the numerical model are $X \times Y \times Z = 90 \times 84 \times 10$ m. Three-dimensional static analysis was conducted to simulate the deformation of the shield tunnel segment during operation. The Mohr–Coulomb [18] constitutive model was used for the geological layer, and a homogeneous elastic body model was used for the lining. The tunnel segment was treated as a whole, made of C60 reinforced concrete with a thickness of 80 cm. SOLID185 solid elements were used to construct the tunnel support and the surrounding rock. Detailed calculation parameters for the surrounding rock and lining structure are listed in Table 1.

Table 1. Calculation parameters of surrounding rock and supporting structure.

Calculation Parameters	Severe	Modulus of Elasticity	Poisson's Ratio	The Angle of Internal Friction	Cohesion
	γ (kN/m^3)	E (GPa)	μ	φ (°)	C (MPa)
Stratum	19	0.045	0.33	20	38
Segment	23.6	36	0.2	/	/

The contact unit was constructed on the contact surface between the segment structure and the stratum structure to simulate the friction effect between the shield tunnel segment and the stratum. The ANSYS APDL 16.0 classical interface is used to apply the contact element, where the inner surface of the stratum is set as the target surface, and the outer surface of the segment is designated as the contact surface. The contact element's interaction model selects the binding contact, ensuring that the segment and the stratum remain in constant contact. Additionally, the normal contact relation algorithm employs the extended Lagrange multiplier method to achieve a realistic contact condition with zero penetration. On this basis, ensuring that the two models do not penetrate each other under the force condition. The interaction model of the contact unit was chosen to bind the contact so that the segment was always in contact with the formation. Meanwhile, the normal contact relationship algorithm was chosen to extend the Lagrange multiplier method to achieve a real contact condition where the penetration was zero. The friction coefficient between contact elements was set to the actual geological friction coefficient, which is 0.4.

The segment and formation model has lateral constraints on the four faces of X = 0, vertical constraints on the four faces of Y = 0, and vertical constraints on the faces of Z = 0 and Z = −4. The ground surface was set as a free boundary, and the lateral stress was determined by the lateral pressure coefficient.

2.1.2. Measuring Point Layout and Working Condition Setting

To analyze the structural deformation of the shield tunnel segment under the cross-section of water pressure and earth pressure. The tunnel cross-section with a longitudinal distance of 5 m was taken as the measuring section. The measuring target was the inner surface of the segment. The structural deformation measuring points were set at intervals of 18 degrees of arc angle, and 20 measuring points were set in total. In the subsequent analysis, the points from the above parts will be used, which are distinguished by different color markers, as shown in Figure 1.

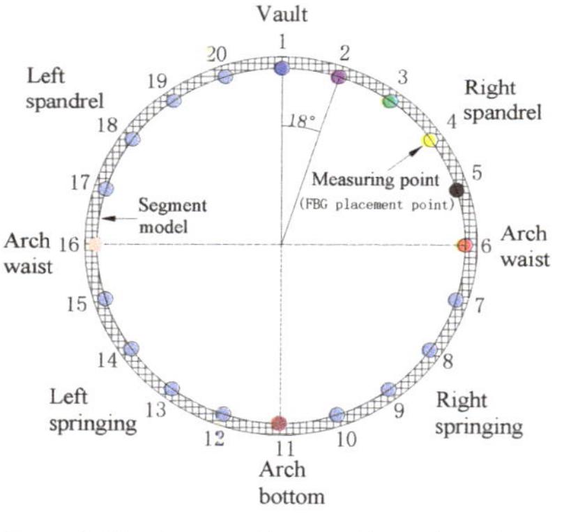

Figure 1. The layout of measuring points for segment deformation of the shield tunnel.

According to the maximum water level of 33.09 m based on practical engineering and the permeability of the covered soil of the tunnel was micro-permeable or weakly permeable. The experiment applied uniformly distributed water pressure (0.33 MPa) and earth pressure (vertical load range of 0.4–2.4 MPa) sequentially. The lateral pressure coefficient of earth pressure was 0.5, and the earth pressure was gradually loaded from 0 MPa to 2.4 MPa with a gradient of 0.4 MPa. 0.33 MPa uniformly distributed water pressure is used to simulate the maximum water level of 33.09 m in the actual project. Earth pressures of 0.4 MPa and 0.8 MPa, respectively, correspond to the minimum overburden thicknesses of 25 m and maximum overburden thicknesses of 38 m.

2.2. *Scale Model Experiment of Underwater Shield Tunnels*

2.2.1. Similar Relationship

Considering the reliability, economy, and flexibility of the model experiment, the geometric similarity ratio of 1:20 and the gravity similarity ratio of 1:1 was selected for the model experiment. According to the three similarity theorems, the similarity relationship of relevant physical and mechanical quantities is determined. The similarity ratio of Poisson's ratio, strain, and friction angle: $C_u = C_\varepsilon = C_\varphi = 1$; the similarity ratio of strength, stress, elastic modulus, and displacement: $C_R = C_\sigma = C_E = C_\delta = 20$.

2.2.2. Preparation of Similar Materials

(1) Segment Model Fabrication

According to the similar ratio mentioned above, a segment model was made with a fixed water-to-gypsum ratio of 1:1.4. The total height of the model is 20 cm (ring height 10 cm, half-ring height 5 cm), the outer diameter is 0.38 m, and the inner diameter is 0.35 m.

After the main body of the shield tunnel segment model is completed, we will conduct a segment joint simulation. For the longitudinal joint form of the segment, the segment joint was simulated through slotting [19], and a slotted joint of a certain depth was opened in the part of the model ring corresponding to the tunnel joint to weaken the flexural stiffness of this part, as shown in Figure 2. The depth of the slotted joints was arranged according to the equivalence principle of unequal positive and negative flexural stiffnesses with the prototype joint. For the segment ring joint form, bolts with a diameter of 4 mm and a length of 20 mm were inserted longitudinally from the tunnel into the inside of the model at the corresponding ring joint position on the model, as shown in Figure 2. The segment rings were spliced, and the shear stiffness between the structural rings is regarded as infinite.

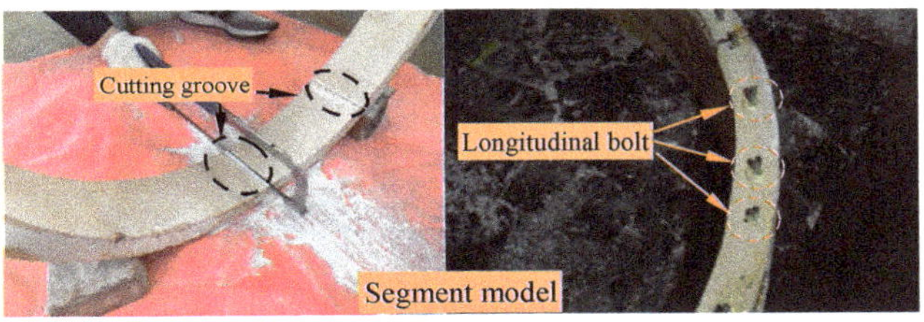

Figure 2. Simulation of segment joint of shield tunnel.

(2) Formation Material Preparation

The experiment's geometric similarity ratio of 1:20 was taken as the basic similarity ratio for the formation of similar material. Based on the research on similar material [20], the mixture of barite powder, fly ash, engine oil, and river sand was used to simulate the prototype formation. The control parameters of the model soil are elastic modulus E, bulk density γ, internal friction angle φ, and cohesion c. The control parameters of the model soil can be calculated according to the similarity ratio, as shown in Table 2.

Table 2. Similarity parameters of model experiment soil.

Mechanical Parameter	Severe γ (kN/m^3)	Modulus of Elasticity E (MPa)	Cohesive Force c (MPa)	Internal Friction Angle φ (°)
Prototype soil	19	45	38	20
Model soil	19	2.25	1.95	20

According to the model soil parameters described in Table 2, combined with the study of a similar stratum model of a shield [21], the weight ratios of barite powder, fly ash, engine oil, and river sand were determined to be 0.25, 0.225, 0.125, and 0.4, respectively.

2.2.3. Test Apparatus

(1) Experiment Device

This experiment required water loading and surrounding rock pressure loading on the segment model. The shield tunnel-stratum similar model test system was used to simulate the surrounding rock pressure loading, as shown in Figure 3. The water pressure loading device was used to simulate the water loading. It mainly consisted of two symmetrically placed water adjustment devices and eight steel strands. By applying torque to the eight steel strands through the adjustment devices, the steel strands were tightened around the segment ring. The water pressure value was confirmed using the tension gauge on the left, and finally, the steel strands were fixed by bolts on the adjustment devices to stabilize the ring towards the water pressure value.

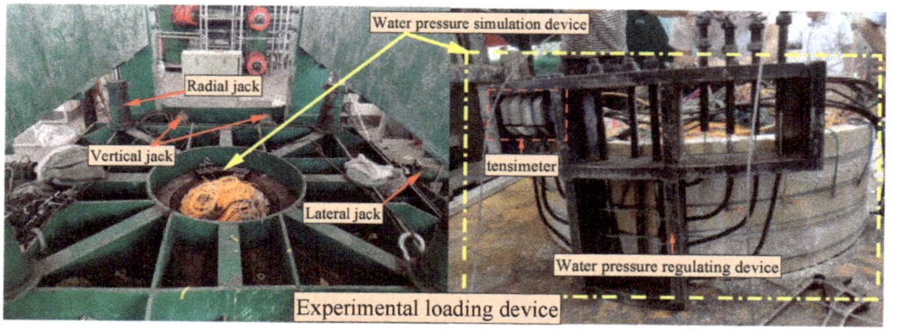

Figure 3. Earth pressure and water pressure loading device.

Due to the advantages of high precision, good stability, high sensitivity, and convenient use, fiber grating is often introduced into shield tunnel deformation measuring [22,23]. To collect the deformation information of the segment model under the combined action of water pressure and surrounding rock pressure, fiber Bragg grating (FBG) was used as a deformation measuring sensor. Select the middle section of the segment model complete ring as the structure deformation sensing object, and evenly arrange FBG sensors on the inner surface of the segment in the circumferential direction. In the similarity relationship, the similarity ratio of the angle is 1:1, ensuring that the arrangement of measuring points in the scale test aligns with that in the simulation test. The length of the elements was selected as 8 cm. The structural deformation measuring points were set at intervals of 18 degrees of arc angle, and 20 FBG sensors were arranged in the circumferential direction, as shown in Figure 1.

(2) Loading Scheme

Place the shield tunnel segment model with the pasted FBG sensing element on the experiment platform, fill the outer gap of the segment model with prepared similar soil, and compact it to fix the model. Radial, vertical, and lateral loads of the segment model were applied by jacks located above the platform and in front, back, left, and right positions.

(3) Experimental Conditions

Based on the maximum water pressure of the actual project and the similarity relationship, the test water pressure is 0.322 MPa. The coefficient of lateral pressure was selected as 0.5. Surrounding rock pressures were applied to the model in three directions, as shown in Figure 3. The vertical surrounding rock pressure of the tunnel was gradually loaded from 0.4 MPa to 6.0 MPa with an increment of 0.4 MPa. The lateral surrounding rock pressure of the tunnel was gradually loaded from 0.2 MPa to 3.0 MPa with an increment of 0.2 MPa. The radial surrounding rock pressure of the tunnel only ensured that there was no radial deformation between the segment and soil, so the radial surrounding rock pressure was maintained at 16.0 MPa. After each level of surrounding rock pressure was loaded, a stable pressure state was maintained for 5 min to record the deformation data of the segment.

3. Results and Discussion

When the surrounding rock pressure reaches 3.6 MPa as applied by the loading device, the similar soil becomes completely compacted, and at this time, the load was fully applied to the surface of the segment model.

To compare and verify the results between the model experiment and numerical simulation, we took the complete earth pressure that the segment was subjected to when the surrounding rock pressure was loaded to 3.6 MPa as the real load of the segment with 0.0 MPa in numerical simulation. The segment is in a deformed state under load in the range of 4.0 to 6.0 MPa, corresponding to a real load on the segment in the range of 0.4 to 2.4 MPa in numerical simulation.

3.1. The Strain of the Segment

Based on the numerical simulation results, the deformation of the tunnel segment structure exhibits obvious characteristics within the loading range from 0 to 2.4 MPa. Especially under the earth pressures of 0.4 MPa and 0.8 MPa, the difference in the effect of water pressure on the deformation of the segment is more significant. Starting from the right arch waist (No. 6), the strain values of each point are plotted in the diagram, following a counterclockwise direction of Figure 1 as the standard. The related deformation characteristics are shown in Figure 4.

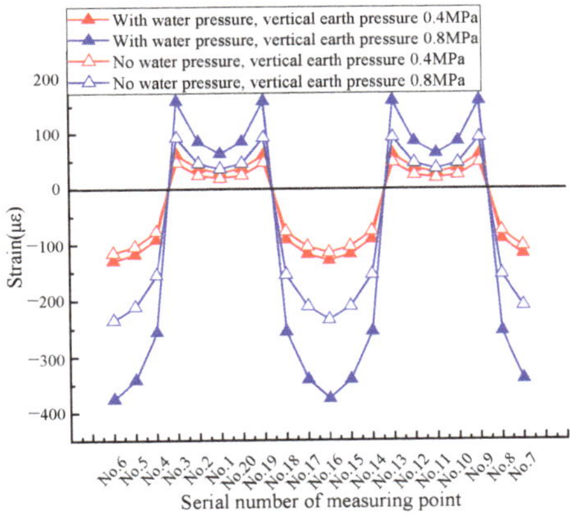

Figure 4. The internal surface strain of the segment with and without water pressure.

Under the condition of earth pressure, the strain of the segment with water pressure is larger than that without water pressure, and the strain difference of the arch crown is 10.0 με, and that of the arch waist is 12 με under 0.4 MPa load. Under the load of 0.8 MPa, the strain difference of the arch crown is 27.6 με, and that of the arch waist is 141.4 με. However, the strain difference of the segment at each measuring point is relatively stable under 0.4 MPa and fluctuates greatly under 0.8 MPa.

According to the analysis above, it can be concluded that the uniform water pressure of 0.33 MPa results in stable strain variation of the segment under a vertical earth pressure of 0.4 MPa, which is beneficial to the stability of the segment structure. However, under an earth pressure of 0.8 MPa, the strain fluctuates greatly, and the advantageous effect of water pressure after this load is reduced. That is, the uniform water pressure of 0.33 MPa has an effective stabilizing effect on the strain of the segment under a vertical load of 0.4 MPa, avoiding structural damage caused by large strain fluctuations of the segment.

Due to the symmetrical deformation of the segment, in this study, the strain information of the 1/4 ring segment under different earth pressures was taken to analyze the deformation characteristics and strain features of the segment under the combined action of uniform water pressure and earth pressure. The load–strain curve is shown in Figure 5.

The tensile strain occurred at the crown, right shoulder 2 and 3 measurement points of the segment, while the compressive strain occurred at the right shoulder 4, 5 and the crown waist measurement points and the strain on the inner surface of the segment continuously increased with the increase of earth pressure. After applying water pressure to the segment, the strain growth rate between various measurement points increased unevenly with the increase of earth pressure, exhibiting a fluctuating trend. Subsequently, the shield tunnel segment experiences a brief and gradual decrease in strain within the loading range. The arch compression strain of the segment decreases from −376.0 με to −344.0 με, occurring in the range of 0.8–1.2 MPa, as shown earlier in Figure 4. While the crown tension strain of the segment decreases from 126.4 με to 84.6 με, occurring in the range of 1.2–1.6 MPa, as shown in the "late" in Figure 4. After the appearance of a slow decrease in the strain at each measurement point on the segment, the strain continues to increase at a growth rate of 21–35% with the increasing earth pressure.

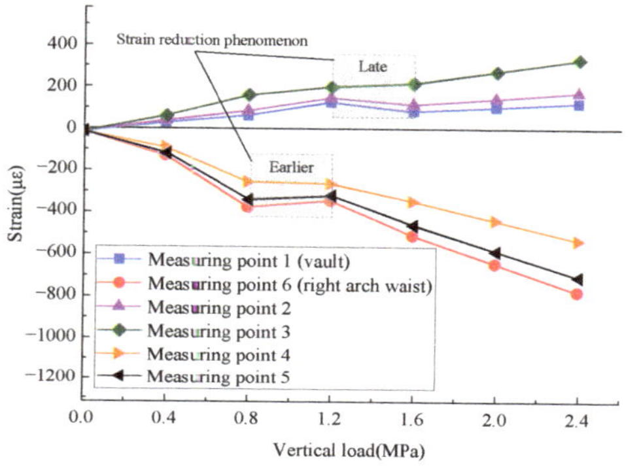

Figure 5. Strain curve of segment structure under load condition.

Therefore, the variations in tensile strain and compressive strain for the segment are not synchronous under the combined action of water pressure and earth pressure. The phenomenon of strain reduction at the arch waist position appears earlier than that at the vault position. Combined with the material characteristics of reinforced concrete, it can be known that the tensile strength of concrete is small, and the tensile side is more prone to crack damage. Therefore, the phenomenon of strain reduction at the arch waist position can be taken as the threshold starting point of a large tensile strain of the segment structure, which will be beneficial to the deformation monitoring and early warning of the underwater shield tunnel segment.

The strain results of the model experiment are shown in Figure 6. It can be seen from Figure 6 that, with a uniformly distributed water pressure of 322 N applied, the maximum tensile strain on the inner surface of the segment model is 43.56 με (at measurement point 17) and the maximum compressive strain on the surface is 47.68 με (at measurement point 7). When the experimental loading device loaded the stratum load of 3.6 MPa, the strain distribution of the segment model showed a more significant change. The maximum compressive strain of the segment was −74.63 με, and the maximum tensile strain was 18.41 με. When the stratum load was 4.0 MPa, the maximum compressive strain of the segment was −177.38 με, and the maximum tensile strain was 86.23 με. When the stratum load was 4.4 MPa, the maximum compressive strain of the segment was −320.81 με, and the maximum tensile strain was 144.38 με. With the increase of the stratum load, the strain distribution of the segment model showed a significant increase in amplitude change. The strain on both sides of the segment arch waist shifted inward, and the offset gradually increased. The strain at the top and waist of the segment shifted outward, and the offset gradually increased. When the earth pressure increases to 5.6 MPa, there were significant numerical changes in the strain distribution of the segment model, which were caused by the material properties of the model itself and the development of internal cracks. At this point, penetrating cracks had appeared in the segment model, resulting in the failure and destruction of the model.

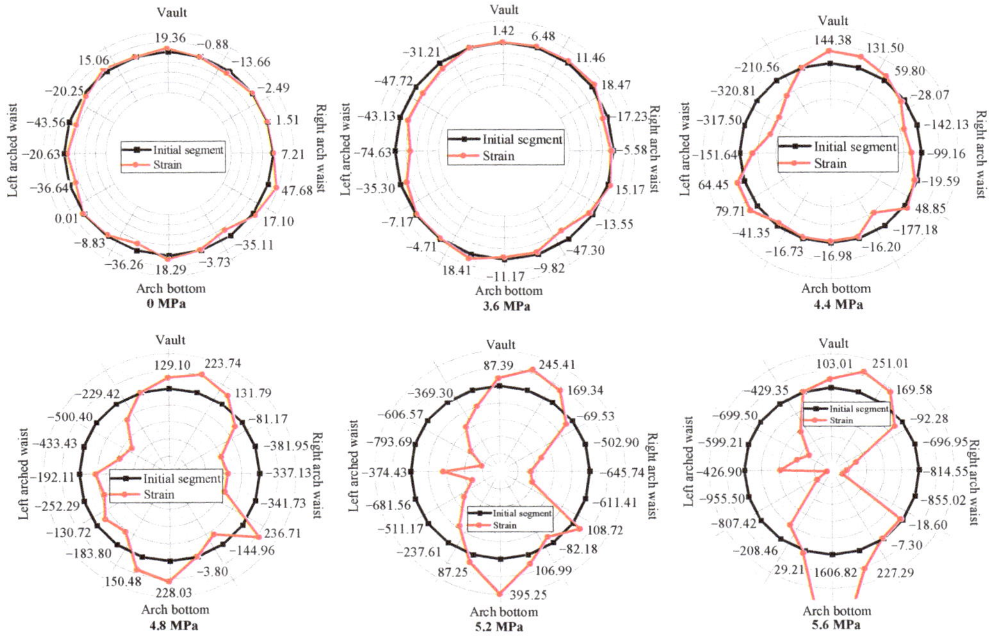

Figure 6. Strain distribution diagram of the segment under soil pressure after applying water pressure.

The strain distribution map of the segment model reveals that the structural strain of the segment increases continuously during the loading process of the surrounding rock pressure. Specifically, tensile strain occurs in the interval between the arch top and arch bottom, while compressive strain takes place within the intervals of the left and right arch waist. The real-time deformation of the segment model exhibits a characteristic feature of inward convergence at the arch top and arch bottom, and outward protrusion at the left and right arch waist.

In summary, two identical conclusions were obtained by comparing the numerical simulation results under the action of water–earth pressure. One is that the strain of the model experiment segment is not synchronized; the strain reduction around the arch waist occurs earlier than around the arch crown. Another is that when the segment structure is subjected to the pressure of the surrounding rock, the tensile strain at the arch top and arch bottom gradually increases, and the structure of the arch top and arch bottom contracts and deforms inward. The compressive strain of the arch waist gradually increases, and the arch waist structures on both sides protrude outward and deform.

3.2. The Curvature of the Segment

In numerical simulations, due to the segment being a homogeneous circular ring, its deformation presents a symmetrical form, and the structural deformation is more likely to occur at the crown and haunch locations of the segment. Therefore, the curvature changes data of the two measurement point locations: the crown and haunch of the segment were selected for analysis. To facilitate data acquisition and calculation of the curvature data, it was assumed that the measurement points and adjacent nodes are on the same circular arc line when calculating the curvature data. The curvature changes curves of the crown and haunch measurement points of the segment under two water pressure conditions are shown in Figure 7.

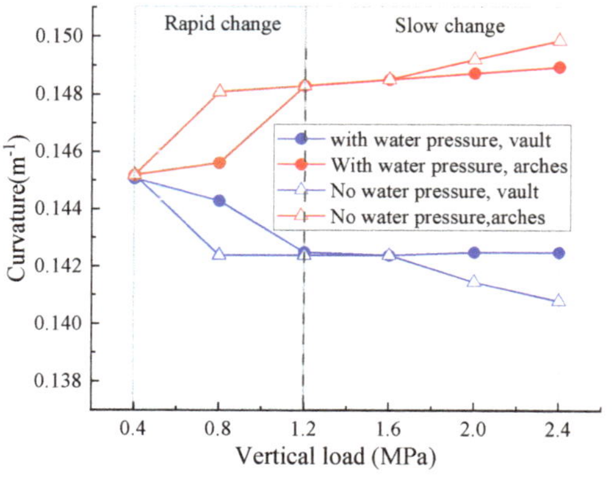

Figure 7. Segment characteristic position curvature changes under load.

Regardless of the presence of water pressure, as the soil pressure increases, the curvature of the crown continuously decreases, while the curvature of the segment waist continuously increases. It can be seen from Figure 7 that when the vertical soil pressure is from 0.4 to 1.2 MPa, the curvature undergoes a rapid increase and decrease stage; when the vertical soil pressure is between 1.2 and 2.4 MPa, the curvature undergoes a slow increase and decrease stage. During the vertical soil pressure of between 1.2 and 2.4 MPa, when there is water pressure, the average rate of change in curvature of the arch crown is 32%, while without water pressure, it is 35%. Regardless of whether there is water pressure or not, the average rate of change in curvature of the arch waist is 39%. During the slow increase and decrease stage of curvature, without water pressure, the average rate of change in curvature of both the arch crown and arch waist is 13%, and with water pressure, the variation range of the curvature of the arch crown and arch waist tends to be close to 0.

In summary, under both conditions of water pressure and no water pressure, the curvature variation of the segment shows a similar changing trend but exhibits different characteristics. Specifically, firstly, without water pressure, the curvature of the arch crown and arch waist positions of the segment exhibit linear changing characteristics. That is, the curvature at the arch crown position first increases rapidly and then gradually increases linearly, while at the arch waist position, the curvature first decreases rapidly and then gradually decreases linearly. Secondly, with water pressure, the curvature at the arch crown position of the segment first increases at a slower speed then increases rapidly, and finally tends to be stable; while the curvature at the arch waist position first decreases at a slower speed, then decreases rapidly, and eventually stabilizes. Thirdly, in the vertical earth pressure range from 0.4 to 1.2 MPa, the amplitude of curvature variation of both the arch crown and arch waist is the same, while in the vertical earth pressure range between 1.2 and 2.4 MPa, the curvature variation amplitude of both the arch crown and arch waist of the segment decrease when there is water pressure.

Regarding the analysis of the model experiment results, since the crown is a common location of deformation in engineering, the curvature of the key position measurement points of the segmental lining model, namely the crown, was selected as an example crown to analyze the deformation law of the shield tunnel segmental lining structure. The load–curvature variation curve of the crown measurement point of the segmental lining model is shown in Figure 8.

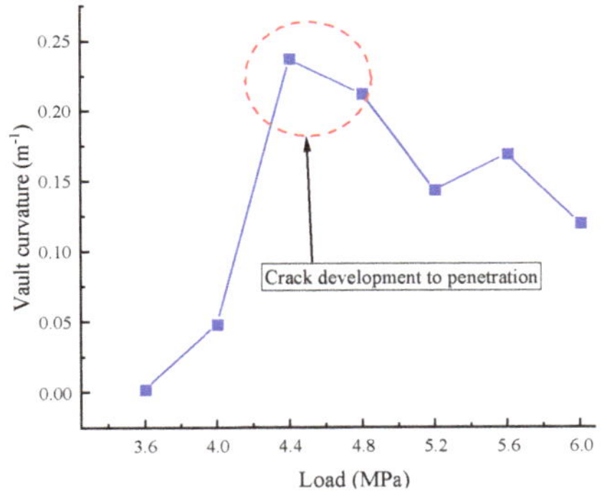

Figure 8. The curve of curvature changes of segment arch crown under model experiment conditions.

In Figure 8, the curvature of measuring point 1 at the arch crown of the shield tunnel segment model sharply increases (with a maximum slope of 0.3) in the load range of 3.6 to 4.4 MPa until the vertical soil pressure reaches 4.4 MPa and a crack pass through, causing the release of structural stress and leading to a sudden drop in the curvature at the arch crown. At the same time, with the development of crack penetration and secondary cracks, the bearing capacity of the segment gradually decreases, and the curvature decreases gradually. Combined with the strain analysis, it can be seen that the curvature can accurately reflect the structural deformation state information of the segment model, and it can be considered that when the curvature increases sharply, cracks may appear in the shield segment.

3.3. The Displacement of the Segment

Through numerical simulation calculation, the displacement change data of segment characteristic position under water pressure and no water pressure are obtained, as shown in Figure 9. Whether water pressure exists or not, the displacement of the vault is larger than that of the arch waist in the whole range of earth pressure, which accords with the deformation law of the tunnel structure. However, the existence of water pressure on the deformation of the segment produced a disturbance, making the segment displacement change. When the shield tunnel is only subjected to earth pressure, the displacement of the segment vault and arch waist is linear with the earth pressure, the displacement of the vault increases with the growth rate of −7.77 mm/MPa, and the displacement of the arch waist increases with the growth rate of 5.51 mm/MPa. When the shield tunnel is subjected to uniform water pressure and earth pressure, the displacement of the vault and the arch waist of the segment fluctuate in the range of 0.4~1.6 MPa earth pressure. When the earth pressure exceeds 1.6 MPa, the displacement of the segment is linear with the earth pressure. The increase rate of the vault displacement is −7.45 mm/MPa, and the displacement of the arch waist is 5.23 mm/MPa. It can be seen that the load-displacement curve at this time is very close to the displacement curve without water pressure.

According to the characteristics of strain variation of the segment, it can be concluded that the displacement of the segment will be disturbed by uniform water pressure, and the deformation fluctuation of the segment will occur under small earth pressure (the displacement decreases with the increased load). Because the displacement fluctuation phenomenon only occurs in the load range where the earth pressure is less than 1.6 MPa, it can be considered that the displacement fluctuation phenomenon caused by water pressure only exists in the small, buried depth (earth pressure is less than 1.6 MPa). Similarly, when the embedded depth of the shield tunnel segment is large (the earth pressure is greater than 1.6 MPa), it can be considered that the displacement of the segment increases linearly with the earth pressure.

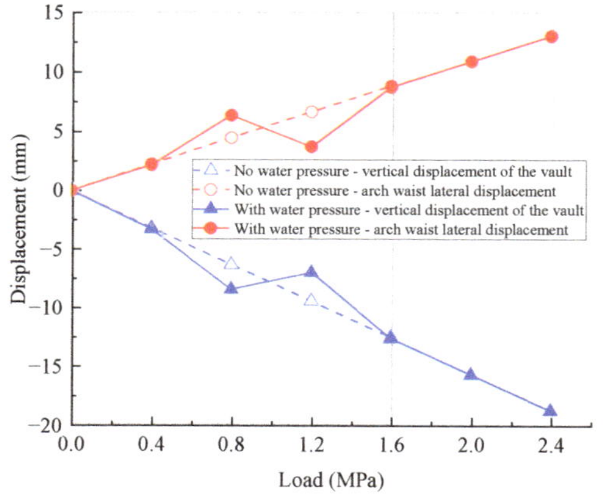

Figure 9. Displacement curve of segment characteristic position.

For the model experiment results, the displacement data at the key locations of the tunnel were taken to draw displacement curves, as shown in Figure 10. Since the displacement gauges were installed after loading water pressure, the state of the segment after loading water pressure was considered as the initial state of displacement, and the displacement was zero.

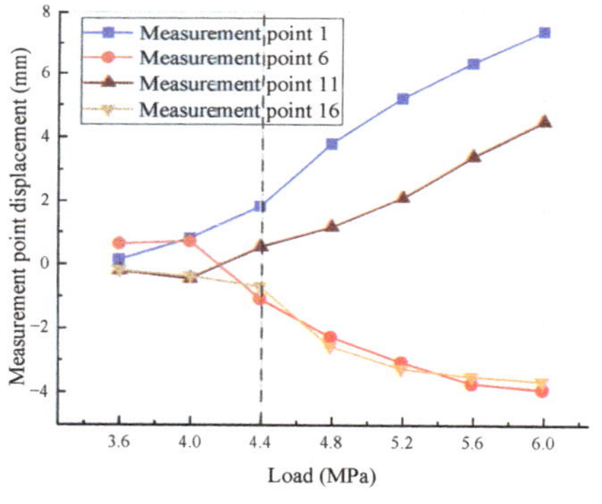

Figure 10. The curve of displacement changes at key positions of the segment under model experiment conditions.

It can be seen from Figure 10 that the shield tunnel segment vault (measuring point 1) and the vault bottom (measuring point 11) both produce inward shrinkage deformation, and the left and right arch waist positions both produce outward expansion deformation. In the load range from 3.6 to 4.0 MPa, the deformation of the segment lining is small and only changes slowly within 0.5 mm. During loading, the arch crown and right arch waist show a trend of inward displacement, while the arch foot and left arch waist show a trend of outward expansion displacement. In the load range from 4.0 to 5.2 MPa, the displacement shows a fluctuating trend. Combined with the displacement results of numerical simulation under the load range from 0 to 0.4 MPa (corresponding to the load range of the model experiment from 3.6 to 4.0 MPa), it can be seen that the uniformly distributed water pressure acting on the segment lining is beneficial for the segment to withstand a certain range of soil pressure. Under the combined action of uniformly distributed water pressure and smaller earth pressure (load range from 3.6 to 4.0 MPa), the segment undergoes asymmetric deformation, with the arch crown and right arch waist (right shoulder area) exhibiting inward contraction deformation, and the arch foot and left arch waist (left shoulder area) exhibiting outward expansion deformation. When uniformly distributed water pressure and larger earth pressure (load range from 4.0 to 6.0 MPa) act together, the strengthening effect produced by water pressure is counteracted by earth pressure, and the deformation of the segment becomes symmetric, with the arch crown and arch foot showing inward displacement and the arch waist showing outward displacement.

4. Conclusions

In this study, the deformation of the shield tunnel segment under two conditions with and without water was numerically simulated with ANSYS 16.0. To verify the numerical simulation results, a 1:20 scale model experiment was conducted using a shield tunnel-geological composite system simulation experiment system to obtain information on the deformation of the segment. The deformation characteristics of the shield tunnel segment underwater were explored from structural strain, curvature, and displacement perspectives by comparing and verifying the numerical simulation and model experiment results. The conclusions are as follows:

(1) A uniformly distributed water pressure of 0.33 MPa had a stabilizing effect on the segment strain under a vertical load of 0.4 MPa under the combined action of water and soil pressure. However, the effect was counteracted after reaching 0.4 MPa. The strain at the arch crown and arch waist gradually increased with the increase in load. Both numerical simulation and model experiment results, in a specific load range (numerical simulation from 1.2 to 1.6 MPa, corresponding to model experiment from 4.8 to 5.2 MPa), showed a sudden decrease followed by the rapid increase in strain at the arch crown. In the model experiment, there was no strain mutation phenomenon at the arch waist, but it showed a continuously increasing trend.
(2) The curvature of the shield tunnel segment changes linearly in the stratum without water pressure. The curvature amplitude of the segment decreases in the stratum as water pressure.
(3) In the stratum with water pressure, the displacement of the segment fluctuates in the range of 0–1.6 MPa (corresponding to 3.6–5.2 MPa in the model experiment), while in the range of 1.6–2.4 MPa (corresponding to 5.2–6.0 MPa in the model experiment. The segment has cracks), the displacement of the segment presents linear change.
(4) Before the appearance of the cracks in the scale model, the displacement of the tunnel segment steadily increases. There are significant fluctuations in the values of strain and curvature when cracks occur, while displacement remains unchanged. The combination of strain, curvature, and displacement of the shield tunnel segment structure can well reflect the structural deformation information.

Author Contributions: Conceptualization, C.H., R.X. and P.X.; Methodology, R.X. and D.Z.; Software, Y.L. (Yufu Li), Z.Z. and Y.L. (Yu Liu); Validation, R.X.; Formal Analysis, Y.L. (Yujun Li); Investigation, Y.L. (Yujun Li); Resources, C.H. and R.X.; Data Curation, Y.L. (Yujun Li), Y.L. (Yufu Li) and Z.Z.; Writing—Original Draft Preparation, R.X., Y.L. (Yujun Li) and D.Z.; Writing—Review and Editing, Y.L. (Yujun Li), D.Z. and R.X.; Visualization, Y.L. (Yujun Li) and Y.L. (Yufu Li); Supervision, R.X., C.H. and C.Y.; Project Administration, C.H., R.X. and P.X.; Funding Acquisition, R.X. and Y.L. (Yujun Li). All authors have read and agreed to the published version of the manuscript.

Funding: This work is funded by the General Fund of Chongqing Natural Science Foundation (No. CSTB2023NSCQ-MSX0960), the State Key Laboratory of Mountain Bridge and Tunnel Engineering Fund Project (No. SKLBT-2109), and the Major Science and Technology Special Project of Xinjiang Uygur Autonomous Region (No. 2020A03003-4), and the Postdoctoral Special Funding Project of Chongqing (No. 01.02.9901212074), and the Research and Innovation Program for Graduate Students in Chongqing (No. CYS240454).

Data Availability Statement: The data presented in this study are available on request from the corresponding author. The data are not publicly available due to the restriction of privacy.

Conflicts of Interest: All authors declare that they have no competing interests.

References

1. Su, D.; Chen, W.; Wang, X.; Huang, M.; Pang, X.; Chen, X. Numerical study on transverse deformation characteristics of shield tunnel subject to local soil loosening. *Undergr. Space* **2022**, *7*, 106–121. [CrossRef]
2. Cao, P.F.; Guo, X.L.; Hu, Y.C.; Xu, T.F.; Zhang, L.C. Research on Longitudinal Deformation Characteristics of Shield Tunnel Segments under Water Level Changes. *Railw. Constr. Technol.* **2021**, *12*, 45–48+86. [CrossRef]
3. Feng, K.; He, C.; Zhou, J.; Zhang, Z. Model Test on Impact of Surrounding Rock Deterioration on Segmental Lining Structure for Underwater Shield Tunnel with Large Cross-Section. *Procedia Environ. Sci.* **2012**, *12*, 891–898. [CrossRef]
4. Feng, K.; He, C.; Qiu, Y.; Zhang, L.; Wang, W.; Xie, H.; Zhang, Y.; Cao, S. Full-scale tests on bending behavior of segmental joints for large underwater shield tunnels. *Tunn. Undergr. Space Technol.* **2018**, *75*, 100–116. [CrossRef]
5. Wang, J.H.; Zhang, W.J.; Guo, X.; Koizumi, A.; Tanaka, H. Mechanism for buckling of shield tunnel linings under hydrostatic pressure. *Tunn. Undergr. Space Technol.* **2015**, *49*, 144–155. [CrossRef]
6. Ye, F.; Gou, C.; Sun, H.; Liu, Y.; Xia, Y.; Zhou, Z. Model test study on effective ratio of segment transverse bending rigidity of shield tunnel. *Tunn. Undergr. Space Technol.* **2014**, *41*, 193–205. [CrossRef]
7. Zhang, Z.; Chen, B.; Li, H.; Zhang, H. The performance of mechanical characteristics and failure mode for tunnel concrete lining structure in water-rich layer. *Tunn. Undergr. Space Technol.* **2022**, *121*, 104335. [CrossRef]
8. Yang, J.-P.; Chen, W.-Z.; Li, M.; Tan, X.-J.; Yu, J. Structural health monitoring and analysis of an underwater TBM tunnel. *Tunn. Undergr. Space Technol.* **2018**, *82*, 235–247. [CrossRef]
9. Arnau, O.; Molins, C. Theoretical and numerical analysis of the three-dimensional response of segmental tunnel linings subjected to localized loads. *Tunn. Undergr. Space Technol.* **2015**, *49*, 384–399. [CrossRef]
10. Cheng, X.; Hong, T.; Lu, Z.; Cheng, X. Characterization of Underlying Twin Shield Tunnels Due to Foundation-Excavation Unloading in Soft Soils. An Experimental and Numerical Study. *Appl. Sci.* **2021**, *11*, 10938. [CrossRef]
11. Liu, X.; Dong, Z.; Bai, Y.; Zhu, Y. Investigation of the structural effect induced by stagger joints in segmental tunnel linings. First results from full-scale ring tests. *Tunn. Undergr. Space Technol.* **2017**, *66*, 1–18. [CrossRef]
12. Cheng, H.; Chen, R.; Wu, H.; Meng, F.; Yi, Y. General solutions for the longitudinal deformation of shield tunnels with multiple discontinuities in strata. *Tunn. Undergr. Space Technol.* **2021**, *107*, 103652. [CrossRef]
13. Zhang, W.; Qi, J.; Zhang, G.; Niu, R.; Zhang, C.; He, L.; Lyu, J. Full-scale experimental study on failure characteristics of the key segment in shield tunnel with super-large cross-section. *Tunn. Undergr. Space Technol.* **2022**, *129*, 104671. [CrossRef]
14. Zhang, L.; Feng, K.; Gou, C.; He, C.; Liang, K.; Zhang, H. Failure tests and bearing performance of prototype segmental linings of shield tunnel under high water pressure. *Tunn. Undergr. Space Technol.* **2019**, *92*, 103053. [CrossRef]
15. Deng, Y.; Yan, W.; Yang, M.; Huang, H. Research progress on influence of environmental waters on static strength of concrete. *Adv. Water Conserv.* **2015**, *35*, 99–104. [CrossRef]
16. Yang, N.; Chen, D.; Peng, G.; Xiao, J.; Xu, T. Mechanical properties of concrete under confining pressure after cyclic loading. *J. Water Conserv. Water Transp. Eng.* **2017**, 89–96. [CrossRef]
17. Sheng, Y.; Cheng, X.; Wang, H. Influences of free water on strength of concrete under dynamic loads. *Concrete* **2018**, *9*, 54–58. [CrossRef]
18. Yang, X.; Zhang, D.; Liu, Z. Numerical Simulations of Longitudinal Settlement of Shield Tunnel under Local Loading. *J. Beijing Univ. Technol.* **2012**, *38*, 1220–1224.
19. Yao, Z.M. Research on Deformation State Sensing of Underwater Shield Tunnel Structure Based on Fiber Bragg Grating Sensing. Master's Thesis, Chongqing Jiaotong University, Chongqing, China, 2022.
20. Liu, C.K.; He, C.; Wang, S.M.; Ma, G.Y.; Guo, W.Q.; Lu, D.Y. Model Test Study on the Effect of Crack Length on the Failure Mode of Shield Tunnel Segment Structure. *J. Cent. South Univ. (Nat. Sci. Ed.)* **2019**, *50*, 1447–1456. [CrossRef]
21. He, C.; Zhang, J.G.; Yang, Z. Experimental Study on Structural Mechanics Characteristics of Segment Lining in Layered Composite Stratum. *J. Geotech. Eng.* **2008**, *10*, 1537–1543.
22. Zhang, J.; Yan, Q.; Li, W.; Su, L.; Sun, M.; Yao, C. Failure analysis of a new-type shield tunnel based on distributed optical fiber sensing technology. *Eng. Fail. Anal.* **2022**, *142*, 106748. [CrossRef]
23. Zhao, F.; Lu, X.; Shi, H.; Liu, B.; Liu, S.; Dai, K.; Fan, Y. Study on Stratified Settlement and Weak Reflectivity Fiber Grating Monitoring of Shield Tunnel Crossing Composite Strata. *Appl. Sci.* **2023**, *13*, 1769. [CrossRef]

buildings

MDPI

Article

Influence Analysis of Material Parameter Uncertainties on the Stability Safety Factor of Concrete Gravity Dams: A Probabilistic Method

Daquan Wang [1], Xiang Lu [2,*], Chengzhi Zheng [1], Ke Gong [3], Litan Pan [1], Liang Pei [2] and Zepeng Zhao [4]

1 Huadian Electric Power Research Institute Co., Ltd., Hangzhou 310058, China; daquan-wang@chder.com (D.W.); chengzhi-zheng@chder.com (C.Z.); litan-pan@chder.com (L.P.)
2 College of Water Resource & Hydropower, Sichuan University, No. 24 South Section 1, Yihuan Road, Chengdu 610065, China; pl_scu@scu.edu.cn
3 Dagu Hydropower Branch of Huadian Xizang Energy Co., Ltd., Shannan 856200, China; gongke407@outlook.com
4 Chengdu Municipal Engineering and Research Design Institute Co., Ltd., Chengdu 610023, China; zepengzhao24@gmail.com
* Correspondence: scu-lx@scu.edu.cn

Abstract: Anti-sliding stability safety is a critical issue that must be given sufficient and widespread attention during the entire lifecycle of gravity dams. The calculated anti-sliding stability safety factor (ASS-SF) is usually compared with the allowable value required by the standards in the traditional method, which ignores the influence of material parameter uncertainties and leads to unreasonable safety evaluation results. Therefore, the nonlinear functional relationship between the stability safety factor (SF) and the random variable parameters is constructed based on the response surface equations, and the distribution types of SF sequences calculated by the Monte Carlo sampling are determined, then a probabilistic stability evaluation method for concrete gravity dams is proposed. Engineering application shows that the calculated SF obeys the normal distribution; the minimum guaranteed rate of different sliding paths in a gravity dam is 86.66%, and the guaranteed rate for the overload safety factor (OSF) is 36.00%. The results imply that a guaranteed rate for the allowable value of the ASS-SF should be provided when making the stability safety evaluation of the dams, especially the OSF. The outcome of this research will advance the understanding of stability evaluation of concrete dams, and reduce the potential risk of sliding instability of concrete dams.

Keywords: concrete gravity dams; anti-sliding stability; safety factor; parameter uncertainties; guaranteed rate

1. Introduction

The failure of gravity dams can cause serious consequences, making dam safety a growing concern. Gravity dams are known to have multiple failure modes, including stress-controlled strength failure, overturning failure, and instability failure. The standards for controlling these failure modes vary depending on the load conditions, meaning that the control values differ under static and dynamic loading conditions. The instability failure is one of the most important failure modes for gravity dams, the control of which under the static loadings is discussed in this study. For the traditional determined analysis model of safety factor (SF) for dams, the calculated SF is a determined value, and the criterion for assessing whether the anti-sliding stability of the structure meets the requirements depends on whether it exceeds the allowable safety factor (ASF) specified in the standard [1,2]. However, the determined SF cannot reflect the inherent uncertainties of material properties or the loads acting on the dam system, thus proposing a probability evaluation method of SFs is of great importance for dam safety.

Safety risk analysis is an important part of dam design, construction, and operation, and has increasingly become a research focus and trend in the field of hydraulic and geological engineering, involving static and dynamic behavior analysis, risk analysis methods, and control standards of dams. For the behavior analysis of dams, Lu et al. [3] studied the impact of spatial variability of concrete material parameters (e.g., tensile strength) on the dynamic response of gravity dams. Ran et al. [4] investigated a risk analysis model for gravity dam safety considering material parameter spatial variability, and their case study showed that the instability failure risk is significantly affected by the parameter spatial variability. Khiavi et al. [5] proposed a new probabilistic model to study the behavior of dams under the action of hydrodynamic waves. Hariri-Ardebili et al. [6] investigated the influence of random field of materials on gravity dams using the linear elastic model under the action of the earthquake. For the risk analysis methods of dams, Li et al. [7] proposed a hybrid reliability assessment method for the arch dams, and then the dam's probability reliability index and non-probability reliability index were investigated. Chen et al. [8] developed a seismic reliability assessment model considering the randomness and time-variability of materials. Wu et al. [9] discussed the suggested SF value for the dam slope stability with a dam height of over 200 m by the determined method. Zhou et al. [10] proposed the absolute and relative stress control indices for the arch dam under the condition of penetrated damage. Han et al. [11] proposed safety control indicators for the earth rock dam (height: 250 m) from the aspects of overall deformation, co-ordinated deformation of the dam body, seepage safety, seismic subsidence safety, structural and material durability, and structural safety. Wang et al. [12] studied a probabilistic stability analysis model for earth rock dams using multivariate adaptive regression splines' soft calculation algorithm, which effectively evaluated the effect of parameter changes on the probability of slope failure. Siacara et al. [13] proposed a reliability analysis method by coupling the analysis software and the reliability solver, and the results indicate that the most dangerous sliding surface of the dam slope based on the minimum SF and reliability index is not consistent. Li and Su [14] put forward a determination method for the deformation safety threshold for gravity dams by discussing the relationship between the water pressure component and the reliability index of stability. However, lots of dams have been built, or are about to be built, in alpine valleys with diverse and discontinuous foundation conditions (i.e., joints, cracks, etc.), leading to uncertainties of material properties in the soil–rock foundation [15]. Neglecting material randomness will result in an unreasonable assessment of the stability safety of the gravity dams under complex geological and operating conditions. For the anti-sliding stability safety factor (ASS-SF), the role of the parameter uncertainties when evaluating the anti-sliding stability can be studied from the perspective of probability theory [16]. However, there is currently no effective research on how to quantify its impact, although some probabilistic methods have been used in the probabilistic and reliability analysis of concrete dams [17–19].

As the height of gravity dams increases, their importance becomes more and more significant. These high concrete gravity dams have complex climates, terrains, and geological environments, with high hydraulic loads, high seepage pressures, and complex combinations of weak rock structure. Gravity dam stability is particularly important under the influence of uncertainties such as material parameters, effects, models, and boundary conditions, and evaluating the stability remains a key scientific and technological challenge that has yet to be fully addressed. Therefore, the objective is to put forward a probabilistic assessment model for the ASS-SF in gravity dams. The research results will help quantify the impact of parameter uncertainty on the ASS-SF in the gravity dams, and assist us in truly evaluating the anti-sliding stability safety margin.

2. Overview of the ASF for Concrete Dams

The operation of dams often has enormous social benefits. Once a dam breaks, it will cause huge disasters, especially now that the dam height is getting higher and the storage capacity is getting larger. The allowable safety factor (ASF) is still one of the important parameters used to judge whether the structure is stable, so a reasonable ASF is critical.

Nowadays, the determination of an ASF in the world is formed based on long-term engineering practice, and scholars around the world have conducted lots of relevant research based on these ASFs, involving the calculation method of the SF, the calibration of new dam-building materials' SF, and the impact of non-linear parameters on the stability, etc., as shown in Table 1. The calculation method significantly affects the stability SF of dams or slopes, but the corresponding ASF is unchangeable, which is related to the structural grade, working conditions, etc. Tables 2 and 3 show the standards for SFs in some countries. It can be found that the ASF varies from country to country, and whether there is material strength parameter test information has an important impact on the ASF, that is, the richness of strength parameter information helps to reduce the uncertainty in the determination of ASF. The current standard in China has no relevant provisions or descriptions on the ASF under the premise of strength parameter test information, so it may be unreasonable to adopt this method due to the existence of parameter uncertainty. However, it is difficult to determine the new ASF, so we can start from the impact of parameter uncertainty on the evaluation results, that is, the discriminating criterion that "As long as the calculated SF is bigger than the ASF, the structure is stable" needs to be adjusted considering the uncertainty of parameters. Therefore, it is necessary to propose a material parameter uncertainties-based probabilistic evaluation method for the ASS-SF in gravity dams, which can help to reasonably evaluate the anti-sliding stability safety margin and avoid potential instability failure hazards.

Table 1. A summary of the safety factors study of the dams and slopes.

No.	Types	Methods	Allowable Safety Factor	Country	Ref.
1	GD	RLEM	Peak sliding factor: 3.0 Residual sliding factor: 1.5 Overturning factor: 1.2 Uplifting factor: 1.2	Canada	[20]
2	GD	RLEM	Usual, no material tests: 3.0 Usual, with material tests/unusual, no material tests: 2.0 Unusual, with material tests: 1.5 Extreme, no material tests: 1.3 Extreme, with material tests: 1.1	Canada	[21]
3	CSGD	RLEM	Structure grades I: 2.9 Structure grades II: 2.6 Structure grades III: 2.2	China	[22]
4	AD	RLEM FEM	2.5	China	[23]
5	GD	RLEM	Usual: 1.5 Unusual: 1.2 Extreme: 1.0	Spain	[24]
6	ED	RLEM	At end of construction: 1.3 At steady state: 1.5 During sudden drawdown: 1.3	Ethiopia	[25]
7	S	RLEM	1.0 and 1.5	Iran	[26]
8	S	RLEM	1.3	China	[27]
9	S	SRE	1.0 and 1.5	Indonesia	[28]
10	S	/	1.0	Korea	[29]
11	S	SRE	1.35	USA	[30]
12	S	/	1.4	India	[31]

Note: GD, gravity dam; AD, arch dam; ED: embankment dam; S: slope; CSGD, cemented sand and gravel dam; RLEM, rigid limit equilibrium method; FEM, finite element method; SRE, strength reduction method.

Table 2. Allowable safety factors without considering the uncertainties in the related standards.

Working Condition	Allowable Safety Factors			
	USBR [32]	**USACE [33]**	**FERC [34]**	**China [35]**
Usual	3.0	2.0	3.0	3.0
Unusual	2.0	1.7	2.0	2.5
Extreme	1.0	1.3	/	2.3

Table 3. Allowable safety factors depending on the a priori information in the related standards.

Working Condition	Allowable Safety Factors					
	CAD [36]		**ANCOLD [37]**		**USACE [38]**	
A Priori Information or Not	**No**	**Test of Strength Parameters**	**Not Well Defined**	**Well Defined**	**Well Defined**	**Ordinary**
Usual	3.0	2.0	3.0	2.0	2.0	1.7
Unusual	2.0	1.5	2.0	1.5	1.5	1.3
Extreme	1.31	1.1	1.5	1.3	1.1	1.1

3. Probabilistic Anti-Sliding Stability Evaluation Methods

3.1. Definition of SF

In civil engineering, the SF K is defined as

$$K = \sum R \Big/ \sum S \tag{1}$$

where R and S are the resistance and loading, respectively.

Usually, the ASS-SF K_S in the gravity dams can be given by

$$K_S = \sum_{i=1}^{n} (f_i{'}\sigma_i + c_i{'})l_i \Big/ \sum_{i=1}^{n} \tau_i l_i \tag{2}$$

where $f_i{'}$ and $c_i{'}$ are the shear strength parameters of the foundation element i on the sliding channel; σ_i and τ_i are the normal, shear stresses of element i, respectively; n is the element number alone the sliding channel; l_i is the length of element i along the direction of the sliding channel.

The overload safety factor (OSF) is another important indicator of dam safety. When the loading increases to the K_H multiple of the designed loading, the deep anti-sliding stability in the foundation reaches a critical state, and the corresponding value is the OSF. In this paper, the commonly used over-water level method is introduced to calculate K_H, as shown in Figure 1 and Equation (3).

$$k_H = \frac{\gamma H_0 \Delta H + 0.5\gamma H_0^2}{0.5\gamma H_0^2} = \frac{2\Delta H + H_0}{H_0} = 2\left(\Delta H \Big/ H_0\right) - 1 \tag{3}$$

K_H is closely related to the instability criteria of the gravity dams on complex foundations. The convergence criteria, feature point displacement mutation criteria, and the run-through of plastic zone criteria are the commonly used criteria in civil engineering. However, there is no unified standard for the specific criteria used to determine the critical instability in the gravity dams on complex foundations. The convergence criterion is selected as the instability criterion of the gravity dams in this paper due to its ability to distinguish the limit state of the gravity dams under the random fluctuation of parameters.

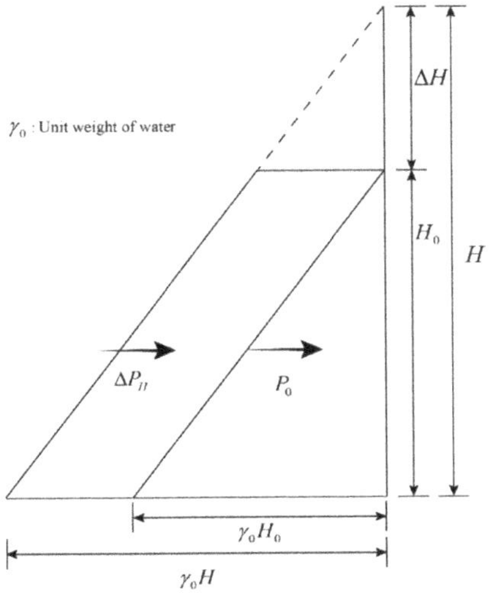

Figure 1. Rectangular overload method of water.

3.2. Probabilistic Evaluation Method for the Safety Factors in Dams

During the operation of a gravity dam, the loading acting on the system and the material parameters have certain fluctuation and randomness, which have a direct relationship with the safety state of the system. However, the functional relationship between the SF and the parameters is difficult to express explicitly. Taking the parameters with uncertainties as random variables (RVs), the functional relationship above can be expressed approximately using the response surface method [39,40], that is, taking the K_S and the K_H as the performance function values and constructing a quadratic response surface equation without cross-terms:

$$K_S \ or \ K_H = a_0 + \sum_{i=1}^{m} b_i x_i + \sum_{i=1}^{m} c_i x_i^2 \tag{4}$$

where m represents the RVs' number; $x_1, x_2, \cdots, x_m$ represents the RV that affects the anti-sliding stability in the gravity dams, such as the elastic modulus, the shear strength index, the upstream water level, etc.; a_0, b_i, c_i are the parameters to be solved.

In this paper, the term "guaranteed rate" of the SF is defined as the probability of being greater than a certain determined value in the SF sequence, which can quantitatively reflect the influence of parameter uncertainties on the SF, thereby making a more reasonable evaluation of dam stability safety. It is usually represented by the cumulative frequency of the calculated value greater than the allowable value. Therefore, if the K_S or the K_H obeys a distribution characteristic with the probability density curve of $f(x)$, the guaranteed rate of SF can be expressed by Equation (5). Figure 2 is the sketch of the guaranteed rate if K_S or K_H obeys normal distribution.

$$P\ (K_S \text{ or } K_H > k) = \int_k^{+\infty} f(x)\mathrm{d}x \tag{5}$$

where P is the SF guaranteed rate; k is the ASF; and $f(x)$ is the SF's probability density distribution function.

Lastly, the flowchart is shown in Figure 3.

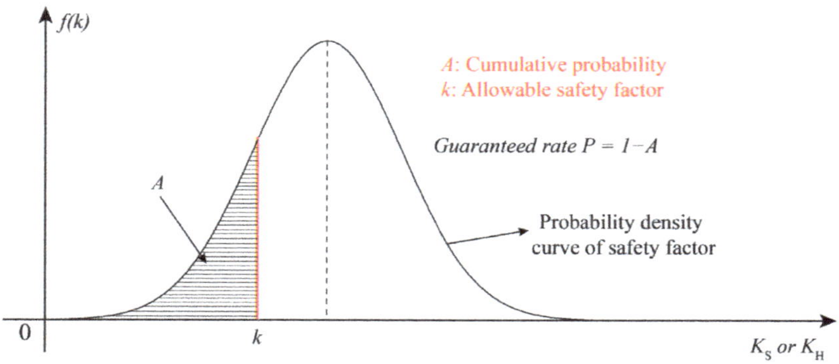

Figure 2. Sketch of the guaranteed rate of the safety factor.

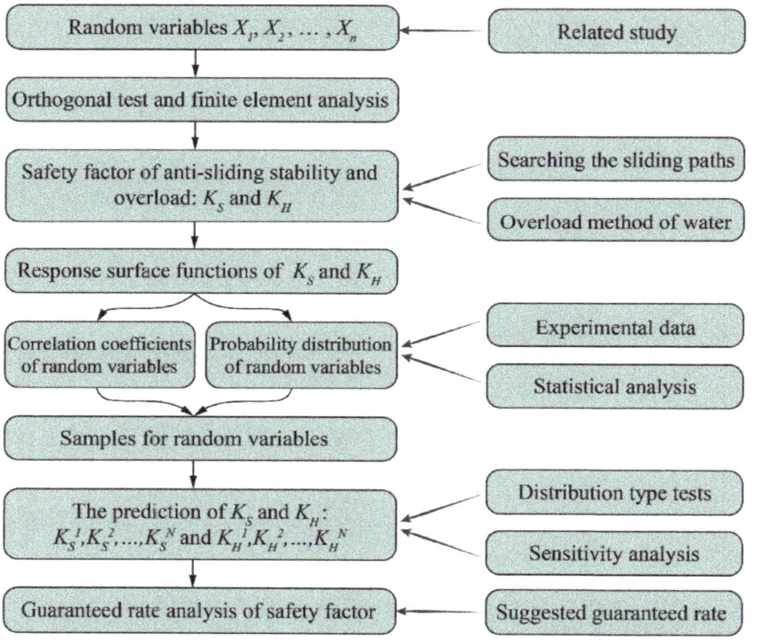

Figure 3. Flow chart of guaranteed rate of safety factors in the gravity dams.

4. Illustrative Example

4.1. Project Specifications

A factual gravity dam with a maximum height of 200 m, crest width of 14 m, and normal storage level (NSL) of 1190 m was adopted in this study to be the case study. In the basalt foundation, there were several groups of joint cracks and two broken zones, A and B, which were detrimental to the dam's stability (Figure 4). A variety of composite slip surfaces were formed due to the multiple joint cracks and broken zones in the rock mass, including four potential deep sliding channels composed of joint fissures and broken zones, and one sliding channel along the foundation surface, as shown in Table 4.

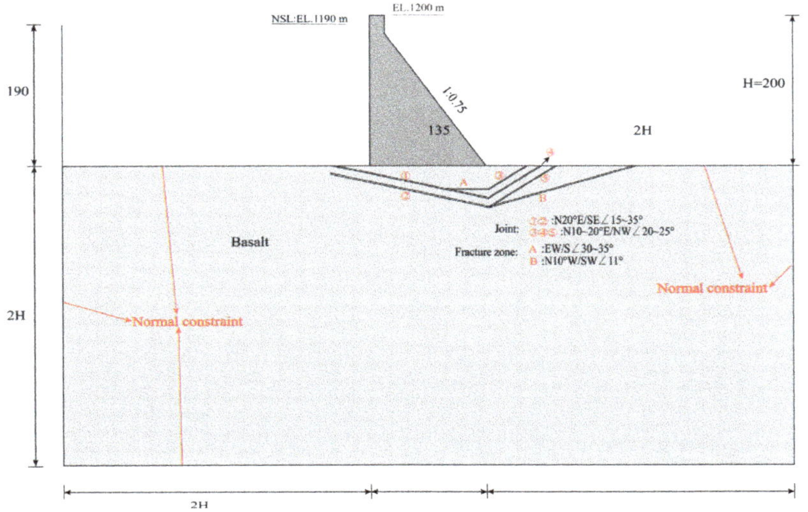

Figure 4. The Sectional drawing of the dam structure (including geological information) (unit: m).

Table 4. Potential sliding channels of the gravity dam.

Code	Component	Safety Factor
Path 1	①, A, ③, ④;	K_{S1}
Path 2	①, ④	K_{S2}
Path 3	②, ⑤	K_{S3}
Path 4	②, B	K_{S4}
Path 5	Foundation surface	K_{S5}

4.2. Dam-Foundation System Model and Parameters

The model was meshed with the ANSYS software (19.0) with the SOLID45 isoparametric elements: the element number of the model was 34,503 and the number of nodes was 47,938. The two sides and the surfaces of the upstream, downstream, and bottom in the foundation were all normal constraints. The calculated loading included the weight of the dam and foundation, upstream water pressure, and uplift pressure.

Dam failure begins with local plastic deformation and gradually progresses to yield, forming a failure path. Considering the mechanical properties and potential failure modes, the Drucker–Prager criterion is introduced in this paper as the criterion for anti-sliding stability failure [41,42]. Relevant research indicates that the stability SF of dams is related to their corresponding failure probability, therefore determining the uncertainty characterization of parameters is crucial [9,43]. Based on the Drucker–Prager model, the test data in a similar gravity dam [44] and the monitoring data, the material parameters, and their statistical characteristics for this dam are shown in Tables 5 and 6.

Table 5. Material parameters of the gravity dam.

Location	E (GPa)	Poisson Ratio	Density (kg/m^3)	c' (MPa)	f'	Note
Dam	28.00	0.20	2400	1.20	1.20	
Foundation	20.00	0.22	2800	2.00	1.40	
Fracture zone	0.50	0.35	2000	0.10	0.60	
Joint	/	/	2800	0.20	0.80	Connectivity rate is 60%

Note: E is the elastic modulus; c' is the cohesive force; and f' is the friction coefficient.

Table 6. Statistical characteristics of Rvs.

Location	Material Parameter	Distribution Type	Mean Value	Variation Coefficient
Dam	E (GPa)	Lognormal	28.00	0.10
	c' (MPa)	Lognormal	1.20	0.25
	f'	Normal	1.20	0.15
Foundation	E (GPa)	Lognormal	20.00	0.15
	c' (MPa)	Lognormal	2.00	0.30
	f'	Normal	1.40	0.20
Fracture zone	E (GPa)	Lognormal	0.50	0.25
	c' (MPa)	Lognormal	0.10	0.35
	f'	Normal	0.60	0.25
Joint	c' (MPa)	Lognormal	0.20	0.35
	f'	Normal	0.80	0.25
Upstream water level (m)		Normal	190.00	0.05

4.3. *Analysis of Safety Factor*

4.3.1. Safety Factor of Anti-Sliding Stability

Reliability is widely recognized as a comprehensive measure of the safety margin across various structures. It can be quantified using the reliability index (RI), which is typically obtained through the Monte Carlo (MC) method [45]. Figure 5 shows the SFs and the RIs of different paths (red line: safety factor; blue line: reliability index). According to the relevant standard in China [46], the ASF $[K]$ is 3.0. The results show that the five paths' SFs are all greater than the ASF, indicating that the anti-sliding stability meets the safety requirements. Additionally, it can be seen that the SF and the RI of the same sliding channel are not completely equivalent. Taking Path 3 and Path 5 as examples, the SF of Path 3 is the largest with a value of 4.62, while its reliability index is not the largest. The reason may be that the influence of the randomness of material parameters is not considered in the calculation of SFs. Therefore, the guaranteed rate of the ASS-SF should be introduced to make a reasonable safety evaluation of the gravity dams.

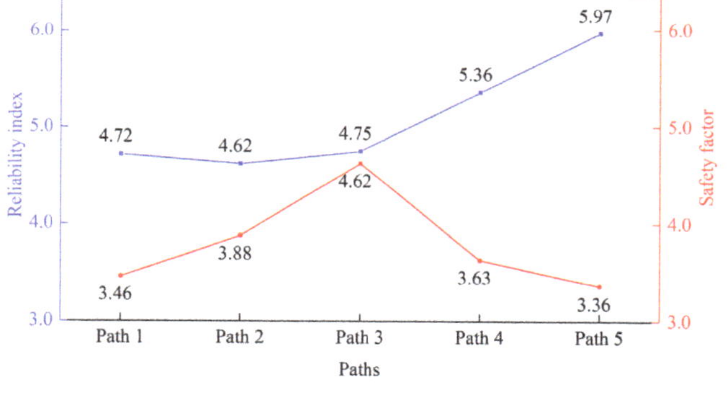

Figure 5. Comparison of the SF and RI in the gravity dams.

4.3.2. Overload Safety Factor

Taking the normal storage level as the beginning of the overloading action, the water pressure of 0.2 times the dam height was selected as the step size, and the overload was carried out step by step until the calculation did not meet the convergence criteria. The development process of the plastic zone at the dam heel under different overloading is shown in Figure 6. The relationship curves among the OSF, the crest displacement, and the cumulative number of iterations are expressed in Figure 7. When $K_H = 1.0$, a small local plastic zone appeared at the dam heel and the upstream side of the foundation joint ①. With the increase in upstream water loading, the area of the plastic zone increased gradually. In detail, the foundation plastic zone at the dam heel developed deeply, the plastic zone in joints ① and ② gradually developed downstream, and the plastic zone was mainly concentrated near joint ①. When $K_H = 2.2$, the plastic strain of joint ① changed rapidly, and the displacement in the dam crest clearly changed (Figure 7a). When $K_H = 3.4$, the numerical simulation did not meet the convergence criterion (Figure 7b), and the dam reached the ultimate bearing state with an OSF of 3.4.

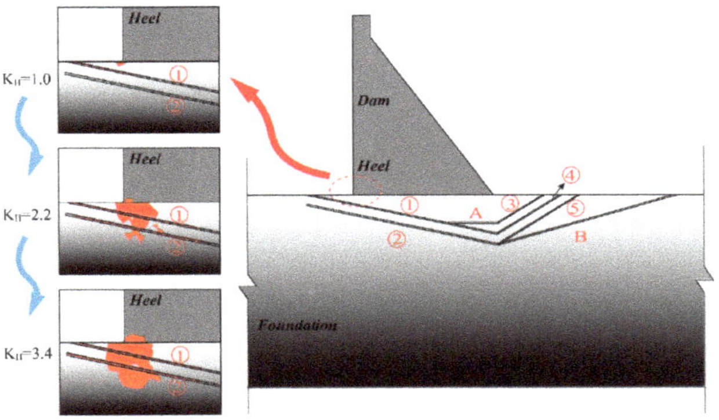

Figure 6. The development process of the plastic zone.

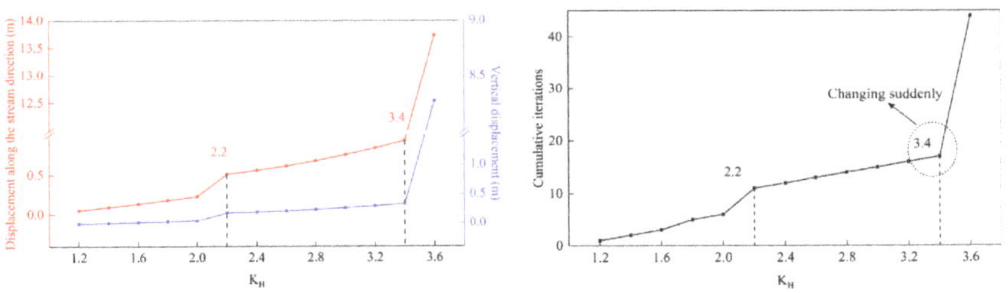

(**a**) Dam crest displacement (**b**) Cumulative number of iterations

Figure 7. Relationship curves among the OSF, the crest displacement, and the cumulative iteration times.

4.4. Probability Analysis of Safety Factor

Because of the small variation coefficients of concrete and rock foundation density, Poisson's ratio, structure size, etc., the randomness of these parameters is not considered in this paper. Therefore, the E, c', f', and upstream water level were chosen as the RVs, and their information is displayed in Table 6.

According to the calculation results, the response surface equation (RSE) sequences are established and solved using the Taguchi design method [47]. The results show that the RSEs' multiple correlation coefficients are above 0.95, implying that the regression effect is of high accuracy.

Based on the RSEs, the MC sampling with a sampling frequency of 10^7 was conducted, and then the distribution characteristics of the SFs were tested using the Kolmogorov–Smirnov (K-S) test, and the results are given in Table 7. The significance of the K–S test is greater than 0.05, indicating that the SF samples obey the normal distribution characteristics. The frequency distribution curve (PDF) and the cumulative distribution curve (CDF) of different SFs are given in Figure 8. Compared with the determined values of the ASS-SF in Figure 5 with the probability of 100%, the SFs calculated by the probabilistic methods have obvious differences. As seen in Figure 8, the guaranteed rate for the SF of 3.36 on the foundation surface (Path 5) is 63.30%, and those of other four Paths 1–4 are between 52.16% and 55.11%. The guaranteed rate for the OSF (3.4) is 36.00%. As seen in Table 7, taking the ASF (3.0) as the evaluation index, the minimum guaranteed rate for the ASS-SF in paths 1–5 is over 86.66%. The results show that it is dangerous when using a determined SF to characterize the stability and safety of the dams. Therefore, a guaranteed rate for the allowable value of ASS-SF should be provided when evaluating the stability safety of the gravity dams, and a corresponding suggested value for the guaranteed rate is not less than 85%. It is noted that the guaranteed rate for the allowable value of the ASS-SF should be determined based on the characteristics of geology, loading, and environment of the gravity dam.

Table 7. Statistical characteristic values and the guaranteed rate of different channel safety factors and overload safety factors sequences.

Statistical Parameters	K_{S1}	K_{S2}	K_{S3}	K_{S4}	K_{S5}	K_H
Mean	3.53	3.96	4.68	3.70	3.56	2.03
Standard deviation	0.41	0.51	0.61	0.41	0.53	0.7
Guaranteed rate	54.90%	53.90%	52.20%	55.10%	63.30%	36.00%
Guaranteed rate for $[K] = 3.0$	89.99%	96.77%	99.72%	95.84%	86.66%	/
Sig. (K-S inspection)	0.13	0.06	0.12	0.11	0.06	0.09

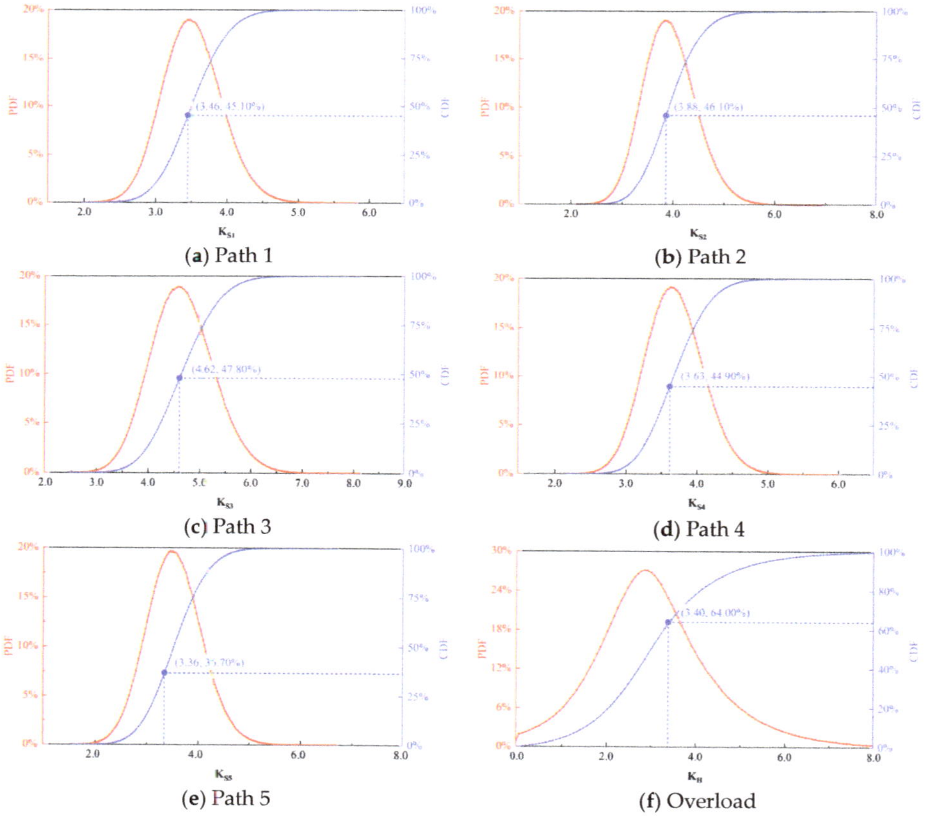

Figure 8. Distribution of safety factors and overload safety factors of different channels.

In addition, the reason for the difference between the ASS-SF and the OSF may be that they are two different methods of evaluating stability safety, and the sensitivity to random variables is different, resulting in different distribution laws.

4.5. Sensitivity Analysis of Material Parameters

4.5.1. Influence of Shear Resistance Parameter Distribution Type on the Guaranteed Rate

Usually, the distribution type of shear strength parameters f' c' generally follows the normal and log-normal distributions in civil engineering. Due to the large dispersion of foundation material parameters, the distribution type of shear strength parameters f' and c' in the dam-foundation system may have different combinations. Four combinations of distribution types for shear strength parameters f' and c' in the gravity dam are expressed in Table 8, and the guaranteed rate of SF is shown in Figure 9. Results show that the guaranteed rate in Case 4 is the lowest, and the anti-sliding stability reliability is also the lowest. Compared with Case 2, which is the suggested combination of shear strength parameters f' and c' in the standard [35], the reliability of Case 4 is more dangerous. Therefore, the determination of the distribution types for shear strength parameters can be carried out based on the shear parameter test values if there is enough geological prospecting data, otherwise it is suggested that the stability reliability of the dam be carried out with the parameter combination of Case 4, that is f' and c', that both follow the log-normal distributions.

Table 8. Combinations of distribution type of shear strength parameters f' and c'.

Parameters	**Distribution**			
	Case 1	**Case 2**	**Case 3**	**Case 4**
c'	Normal	Log-normal	Normal	Log-normal
f'	Normal	Normal	Log-normal	Log-normal

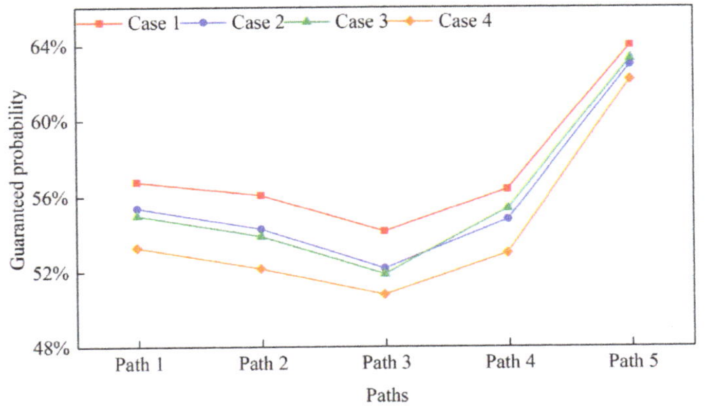

Figure 9. Relationship between combinations of f' and c' of each path and the safety guaranteed rate.

4.5.2. Influence of Correlation of Shear Resistance Parameters on the Guaranteed Rate

Studies have shown that the shear resistance parameters of c' and f' in rock mass have a certain correlation [48]. Based on the four combinations of distribution types in Table 6, the influence of correlation on the guaranteed rate for SFs in the dam-foundation system is discussed and the results are shown in Figure 10. It was found that the guaranteed rate for SFs in paths 1–5 decreases with the increase in correlation coefficient between f' and c'. The reason may be that the increase in the correlation coefficient makes the sampling values more centralized (Figure 11), thereby making the SF sequences divergent. Taking the correlation coefficients 0.9 and −0.9 as examples, the differences among the guaranteed rates under Cases 1–4 are between 3.69% and 18.78%, implying that the influence of the correlation between f' and c' on stability safety evaluation cannot be ignored.

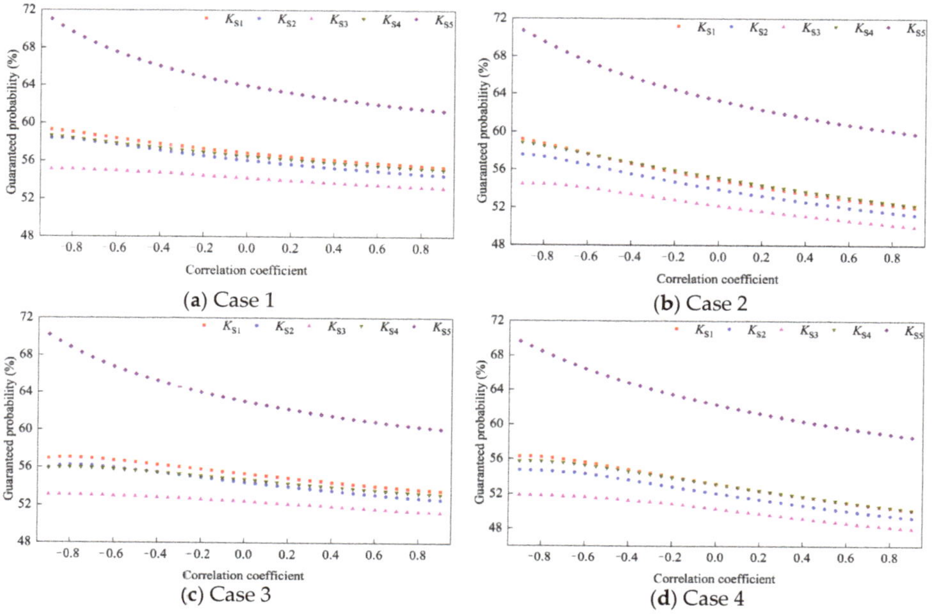

Figure 10. The influence of the correlation coefficient on the guaranteed rate.

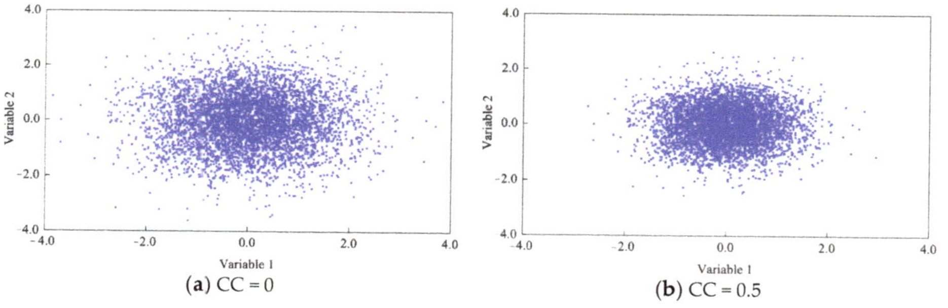

Figure 11. The sampling value with correlation coefficients (CC) of 0 and 0.5.

5. Discussion

5.1. Difference of Discrimination Criteria for Anti-Sliding Stability

From the above research, it can be seen that, under the influence of parameter uncertainty, if the traditional discrimination criteria are used to evaluate the stability safety, the probability level of evaluation results is not 100%. Taking Path 1 in Table 4 as an example, the determined value of the SF is 3.46. For the traditional method, the status of Path 1 is stable due to 3.46 being greater than 3.0, with a potential probability level of 100%. However, the SF is subject to the normal distribution with an average value of 3.53 and a standard deviation of 0.41 after considering the uncertainties of parameters (as shown in Table 7), and the probability level that the SF is bigger than 3.0 is 89.99%, not 100%. What is more, the OSF is a factor used to reflect the capacity of dams to bear overloads. As shown in Figure 7, the OSF is 3.4, but the actual guaranteed rate that can meet the overload capacity is only 36%, indicating that the probability of this dam having an overload capacity of 3.4 times is only 36%, which is why traditional methods are not reasonable. Similarly, the arch dam abutment stability assessment based on SFs also has such problems, and the promotion and application results of an arch dam are displayed in Figure 12. It was found that the guaranteed rate that the SF of dam abutment stability on the banks is bigger than the ASF (3.5) was 99.99%, but the guaranteed rate of the OSF over 6.4 obtained by the traditional method was only 47.12%.

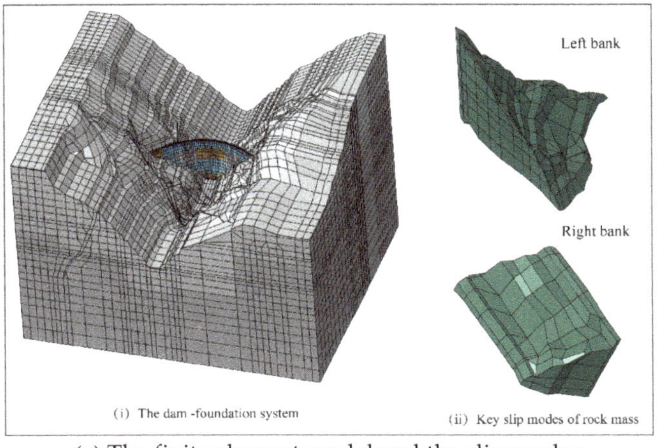

(**a**) The finite element model and the slip modes

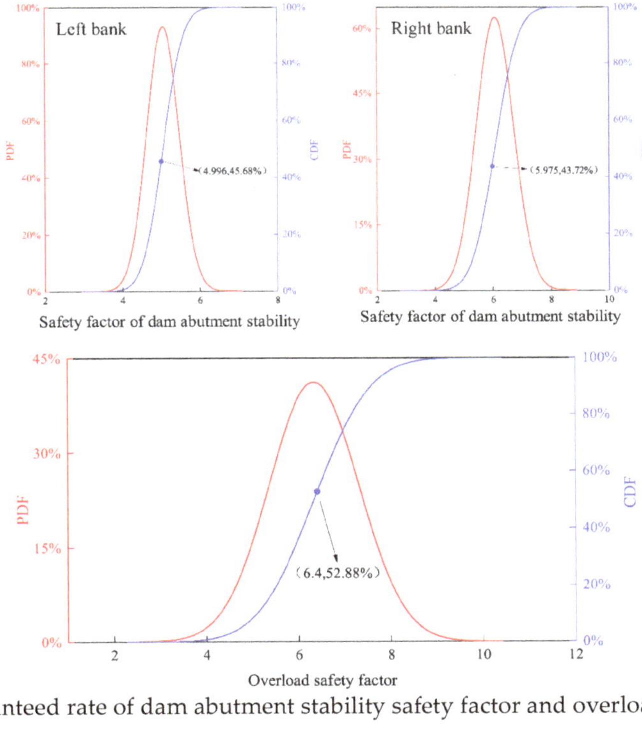

(**b**) The guaranteed rate of dam abutment stability safety factor and overload safety factor

Figure 12. Application results of abutment stability in an arch dam.

Therefore, the method studied in this article can be extended to the stability problems of different types of dams, such as the stability of gravity dams, stability of arch dam abutments, and dam slope stability. The probabilistic methods for SF provide important theoretical support for setting reasonable stability thresholds and also help to clarify the ability of dams to resist extreme environments, especially with the significantly increased probability of extreme environments occurring globally.

5.2. Future Research Directions

It should be noted that the guaranteed rate of using the ASF to evaluate the stability safety in gravity dams and the abutment stability in arch dams is affected by the parameter uncertainty. Maybe these guaranteed rates generally exceed 85%, although there is no significant impact on whether the structure is in a safe state. However, if a specific guaranteed rate can be indicated during the evaluation, it will make the structural safety assessment more reasonable. For the OSF, the guaranteed rate (e.g., 36% for the gravity dam, and 47.12% for the arch dam) is much lower than the stable SF. If the OSF is directly regarded as the overload capacity of the dam, it may cause dam failure in extreme cases. Combined with the research results above, a guaranteed rate of 85% is suggested for the overload capacity assessment of dams. As shown in Table 9, the OSF with a guaranteed rate of 85% is smaller than that obtained by traditional methods, with a decrease rate even approaching 50%, indicating that the OSF obtained by traditional methods cannot reflect the true overload capacity of the dam under extreme conditions. However. Related studies have shown that the ASF for dam slope stability should vary depending on the dam height [9]. Similarly, if we want to extend the content of this study to all gravity dams, a lot of research is still needed, such as the sequence law of ASS-SFs for gravity dams under different dam heights, working conditions, and foundation materials. The model proposed in this paper cannot be directly applied to failure modes such as dam overtopping that do not use SFs as safety indicators, and further research is needed.

Table 9. The overload safety factor for dams with a guaranteed rate of 85%.

Terms	Gravity Dams	Arch Dams
Overload safety factor	3.4	6.4
Guaranteed rate	36%	47.12%
Overload safety factor with a guaranteed rate of 85%	1.82	5.32

6. Conclusions

(1) Taking the uncertain parameters as random variables, the sequences of safety factors and their distribution types were obtained using the RSEs and MC method, respectively, and a probabilistic evaluation method for the ASS-SF in the gravity dams was proposed.

(2) The case study shows that both the ASS-SF and the overload safety factor in gravity dams obeyed the normal distribution, considering the uncertainties of parameters. The guaranteed rates for the ASS-SF and the OSF were 86.66% and 36.00%, respectively. The distribution type combinations and the correlation between f' and c' had a certain impact on the guaranteed rate of the K_s in the gravity dams.

(3) It is indicated that the probability that a determined and unique safety factor that is greater than the ASF during the safety evaluation of concrete dams may not necessarily be 100%. The impact of material parameter uncertainty is not unique but rather depends on the type of dam, failure mode, and material characteristics.

Additionally, the guaranteed rate of the ASS-SF of gravity dams or arch dams is greatly affected by the randomness of parameters. This article only discusses one gravity dam and one arch dam, which is not sufficient to represent all concrete dams. Therefore, specific concrete gravity dam projects need to be discussed. In addition, the guaranteed rate proposed in this article is only a suggested value, and further work is needed based on a large amount of in-depth research.

Author Contributions: Conceptualization, D.W. and X.L.; methodology, D.W., X.L., C.Z., K.G., L.P. (Litan Pan) and Z.Z.; software, C.Z., K.G., L.P. (Litan Pan), L.P. (Liang Pei) and Z.Z.; formal analysis, X.L., K.G., L.P. (Litan Pan) and Z.Z.; investigation, D.W.; resources, C.Z. and L.P. (Litan Pan); data curation, X.L., K.G. and Z.Z.; writing—original draft, D.W.; writing—review and editing, X.L. and L.P. (Liang Pei). All authors have read and agreed to the published version of the manuscript.

Funding: This research was supported by the National Natural Science Foundation of China (No. 52309162) and the Fundamental Research Funds for the Central Universities (YJ202287).

Data Availability Statement: The data presented in this study are available on request from the corresponding author.

Conflicts of Interest: Authors Daquan Wang, Chengzhi Zheng, and Litan Pan were employed by the company Huadian Electric Power Research Institute Co., Ltd., Ke Gong was employed by the company Dagu Hydropower Branch of Huadian Xizang Energy Co., Ltd., and Zepeng Zhao was employed by the company Chengdu Municipal Engineering and Research Design Institute Co., Ltd. The remaining authors declare that the research was conducted in the absence of any commercial or financial relationships that could be construed as potential conflicts of interest.

References

1. Tao, Z.; Liu, Y.; Yang, Q.; Wang, S. Study on the nonlinear deformation and failure mechanism of a high arch dam and foundation based on geomechanical model test. *Eng. Struct.* **2020**, *207*, 110287. [CrossRef]
2. Shi, C.; Yang, W.; Chu, W.; Shen, J.; Kong, Y. Study of anti-sliding stability of a dam foundation based on the fracture flow method with 3D discrete element code. *Energies* **2017**, *10*, 1544. [CrossRef]
3. Lu, X.; Wu, Z.; Pei, L.; He, K.; Chen, J.; Li, Z.; Yang, Z. Effect of the spatial variability of strength parameters on the dynamic damage characteristics of gravity dams. *Eng. Struct.* **2019**, *183*, 281–289. [CrossRef]
4. Ran, C.; Zhou, Z.; Pei, L.; Lu, X.; Gong, B.; He, K. A method for evaluating systematic risk in dams with random field theory. *Appl. Sci.* **2024**, *14*, 4349. [CrossRef]
5. Khiavi, M.; Ferdousi, A.; Khiavi, A. A probabilistic model for evaluation of the dynamic behavior of a concrete gravity dam considering the fluid-structure interaction. *Adv. Civ. Eng.* **2023**, *2023*, 9927608. [CrossRef]
6. Hariri-Ardebili, M.A.; Seyed-Kolbadi, S.M.; Saouma, V.E.; Salamon, J.; Rajagopalan, B. Random finite element method for the seismic analysis of gravity dams. *Eng. Struct.* **2018**, *171*, 405–420. [CrossRef]
7. Li, Y.; Zhao, E.; Zhang, J.; Shao, C.; Li, Z. Hybrid reliability evaluation of arch dam during long-term service with multi-dimensional parallelepiped convex model. *Eng. Fail. Anal.* **2024**, *157*, 107937. [CrossRef]
8. Chen, J.; Jia, Q.; Xu, Q.; Fan, S.; Liu, P. The PDEM-based time-varying dynamic reliability analysis method for a concrete dam subjected to earthquake. *Structures* **2021**, *33*, 2964–2973. [CrossRef]
9. Wu, Z.Y.; Chen, C.; Lu, X.; Pei, L.; Zhang, L.M. Discussion on the allowable safety factor of slope stability for high rockfill dams in China. *Eng. Geol.* **2020**, *272*, 105666. [CrossRef]
10. Zhou, X.; Xu, Q.; Chen, S. Penetrated damage effects and safety control of arch dam (III): Early warning index of dam body. *Eng. J. Wuhan Univ.* **2023**, *56*, 1033–1040. [CrossRef]
11. Han, X.; Zhao, P.; Mao, Z.; Yang, C. Research on safety control indicators for the design of Dashixia extra high earth rock dam in Xinjiang. *Water Resour. Plan. Des.* **2022**, *3*, 89–95. [CrossRef]
12. Wang, L.; Wu, C.; Liu, H.; Mei, G.; Zhang, W. Probabilistic stability analysis of earth dam slope under transient seepage using multivariate adaptive regression splines. *Bull. Eng. Geol. Environ.* **2020**, *79*, 2763–2775. [CrossRef]
13. Siacara, A.T.; Napa-García, G.F.; Beck, A.T.; Futai, M.M. Reliability analysis of earth dams using direct coupling. *J. Rock Mech. Geotech. Eng.* **2020**, *12*, 366–380. [CrossRef]
14. Li, J.; Su, H. Anti-sliding stability reliability-based approach determining the early-warning value of deformation safety for gravity dam. *HydroSci. Eng.* **2023**, *1*, 34–42. [CrossRef]
15. Gao, J.; Han, X.; Han, W.; Dang, F.; Ren, J.; Xue, H.; Fang, J. Research on the slip deformation characteristics and improvement measures of concrete-faced rockfill dams on dam foundations with large dip angles. *Sci. Rep.* **2024**, *14*, 8665. [CrossRef] [PubMed]
16. Xu, B.; Wang, S. Sensitivity analysis of factors affecting gravity dam anti-sliding stability along a foundation surface using Sobol method. *Water Sci. Eng.* **2023**, *16*, 399–407. [CrossRef]
17. Hariri-Ardebili, M.A.; Saouma, V.E. Probabilistic seismic demand model and optimal intensity measure for concrete dams. *Struct. Saf.* **2016**, *59*, 67–85. [CrossRef]
18. Pouraminian, M.; Pourbakhshian, S.; Noroozinejad Farsangi, E. Reliability assessment and sensitivity analysis of concrete gravity dams by considering uncertainty in reservoir water levels and dam body materials. *Civ. Environ. Eng. Rep.* **2020**, *1*, 1–17. [CrossRef]

19. Hariri-Ardebili, M.A. Quantifying modeling uncertainties in seismic analysis of dams: Insights from an international benchmark study. *Earthq. Eng. Struct. Dyn.* **2023**, *53*, 1168–1194. [CrossRef]
20. Leclerc, M.; Le'ger, P.; Tinawi, R. Computer aided stability analysis of gravity dams-CADAM. *Adv. Eng. Softw.* **2003**, *34*, 403–420. [CrossRef]
21. Cordier, M.; Léger, P. Structural stability of gravity dams: A progressive assessment considering uncertainties in shear strength parameters. *Georisk Assess. Manag. Risk Eng. Syst. Geohazards* **2018**, *12*, 109–122. [CrossRef]
22. Hao, N.; Li, X.; Li, Y.; Jia, S.; Gao, L. A novel reliability-based method of calibrating safety factor: Application to the cemented sand and gravel dams. *Eng. Geol.* **2022**, *306*, 106719. [CrossRef]
23. Bao, T.; Xu, B.; Zheng, X. Hybrid method of limit equilibrium and finite element internal force for analysis of arch dam stability against sliding. *Sci. China Technol. Sci.* **2011**, *54*, 793–798. [CrossRef]
24. Cabrera, M.; Senent, S.; Rafael Jimenez, O. Sliding stability analysis of gravity dam founded on rock mass with impersistent joint set. *Can. Geotech. J.* **2015**, *52*, 1668–1684. [CrossRef]
25. Shole, D.; Belayneh, M. The effect of side slope and clay core shape on the stability of embankment dam: Southern Ethiopia. *Int. J. Environ. Sci. Technol.* **2019**, *16*, 5871–5880. [CrossRef]
26. Johari, A.; Momeni, M.; Javadi, A. An analytical solution for reliability of rock slopes using jointly distributed random variables method. *Iran. J. Sci. Technol. Trans. Civ. Eng.* **2015**, *39*, 351–363. [CrossRef]
27. Wang, S.; Ni, P. Application of block theory modeling on spatial block topological identification to rock slope stability analysis. *Int. J. Comput. Methods* **2014**, *11*, 1350044. [CrossRef]
28. Rochim, A.; Pratikso. Characterization of landslide and its earthwork solutions. *J. Phys. Conf. Ser.* **2020**, *1517*, 012035. [CrossRef]
29. Kim, K.; Park, H.; Lee, S.; Woo, I. Geographic Information System (GIS) based stability analysis of rock cut slopes. *Geosci. J.* **2004**, *8*, 391–400. [CrossRef]
30. Lazarte, C.; Robinson, H.; Gomez, J.; Baxter, A.; Cadden, A.; Berg, R.; Berg, R. *Geotechnical Engineering Circular No. 7 Soil Nail Walls Reference Manual*; National Highway Institute (US): Arlington, VA, USA, 2015.
31. *RDSO Guidelines for Earthwork in Railway Projects*; Guidelines no. GE: G-1; Ministry of Railways, Govt. of India, Geotechnical Engineering Directorate, Research Designs and Standards Organization: Lucknow, India, 2003.
32. *USBR Design of Gravity Dams*; United States Bureau of Reclamation: Denver, CO, USA, 1976.
33. *USACE Gravity DamDesign*; USACE: Washington, DC, USA, 1995.
34. *FERC Engineering Guidelines for the Evaluation of Hydropower Projects*; Federal Energy Regulatory Commission: Washington, DC, USA, 2002.
35. Ministry of Water Resources of the People's Republic of China. *Chinese Code: Design Specification for Concrete Gravity Dams*; China Water & Power Press: Beijing, China, 2018.
36. *CDA Dam Safety Guidelines*; Canadian Dam Association: Edmonton, AB, Canada, 2007.
37. *ANCOLD Guidelines on Design Criteria for Concrete Gravity Dams*; Australian National Committee on Large Dams: Hobart, Australia, 2013.
38. *USACE Stability Analysis of Concrete Structures*; USACE: Washington, DC, USA, 2005.
39. Li, Z.; Lu, D.; Gao, X. Optimization of mixture proportions by statistical experimental design using response surface method—A review. *J. Build. Eng.* **2021**, *36*, 102101. [CrossRef]
40. Pouraminian, M.; Ekranejad, H. Reliability analysis of concrete arch dam under stage construction and hydrostatic pressure by MCS and RS methods. *Sharif Civ. Eng. J.* **2021**, *37*, 135–145. [CrossRef]
41. Liu, J.; Luan, M.; Xu, C.; Wang, J.; Yuan, F. Study on parametric characters of Drucker-Prager criterion. *Chin. J. Rock Mech. Eng.* **2006**, *25*, 4010–4015.
42. Wang, X.; Chen, M.; Chang, X.; Zhou, W.; Yuan, Z. Studies of application of Drucker-Prager yield criteria to stability analysis. *Rock Soil Mech.* **2009**, *30*, 3733–3738.
43. Altarejos-García, L.; Escuder-Bueno, I.; Serrano-Lombillo, A.; Gómez de Membrillera-Ortuño, M. Methodology for estimating the probability of failure by sliding in concrete gravity dams in the context of risk analysis. *Struct. Saf.* **2012**, *36–37*, 1–13. [CrossRef]
44. PowerChina Chengdu Engineering Corporation Limited (PCECL). *Special Report of Determination of Dam Foundation Surface and Review of Anti-Sliding Stability of the Gravity Dam*; PCECL: Chengdu, China, 2010.
45. Pei, L.; Chen, C.; He, K.; Lu, X. System reliability of a gravity dam-foundation system using Bayesian networks. *Reliab. Eng. Syst. Saf.* **2022**, *218*, 108178. [CrossRef]
46. *DL5108-1999*; Design Specification for Concrete Gravity Dams. State Economic and Trade Commission of the People's Republic of China (SETC): Beijing, China, 1999.
47. Tai, S.; Sun, Y.; He, J. *Applied Mathematical Statistic*; Wuhan University Press: Wuhan, China, 2005; pp. 204–231.
48. Yuan, Y.; Guo, Q.; Zhou, Z.; Wu, Z.; Chen, J.; Yao, F. Back analysis of material parameters of high core rockfill dam considering parameters correlation. *Rock Soil Mech.* **2017**, *38*, 463–470. [CrossRef]

buildings

MDPI

Article

Time-Dependent Reliability Analysis of Anhydrite Rock Tunnels under Swelling Conditions: A Study on Stress, Deformation, and Engineering Solutions

Zihui Zan [1], Ran Wang [1], Yunfeng Zhao [2], Jianxun Wu [3] and Zhenkun Hou [4,*]

1 Guangzhou Metro Design & Research Institute Co., Ltd., Guangzhou 510010, China
2 School of Urban Construction, Yangtze University, Jingzhou 434023, China; zhaoyf@yangtzeu.edu.cn
3 East China Electric Power Design Institute Co., Ltd., China Power Engineering Consulting Group, Shanghai 200063, China
4 School of Civil and Transportation Engineering, Guangdong University of Technology, Guangzhou 510006, China
* Correspondence: zhenkunhoucqu@163.com

Abstract: This study presents an analytical approach for evaluating the reliability of anhydrite rock tunnels, focusing on their characteristic swelling behavior. Anhydrite rocks, prone to significant expansion upon moisture exposure, pose a challenge in tunnel construction, potentially leading to structural issues such as floor heave and lining damage. To address this, this research develops an elastic swelling analytical solution based on humidity stress field theory, enabling the assessment of time-dependent stress and deformation changes in anhydrite tunnels. The solution's applicability is demonstrated through its application to the Lirang tunnel. The investigation into the effects of support pressure, swelling time, and reserved deformation on tunnel reliability reveals that circumferential stress at the tunnel wall increases by 13.94% and 21.86% for swelling periods of 30 and 365 days, respectively. Similarly, radial displacement escalates by 22.97% and 35.93% over these periods, highlighting the significant impact of swelling behavior. Using a spreadsheet-based First Order Reliability Method (FORM) for analysis, this study finds that the original design of the Lirang tunnel did not meet the desired reliability standards under swelling conditions. However, strategic adjustments in construction variables, such as increasing support pressure to 1.2 MPa or enhancing reserved deformation to 59 mm, elevated the tunnel's reliability to meet safety requirements. This research provides a vital framework for assessing and enhancing the reliability of anhydrite rock tunnels, considering the long-term effects of swelling. It underscores the importance of incorporating swelling behavior in the design and construction of tunnels in anhydrite rock formations, offering valuable insights for optimizing tunnel stability in such challenging geological conditions.

Keywords: anhydrite rock; swelling behavior; elastic swelling analytical solution; tunnel reliability

1. Introduction

Anhydrite rock, composed mainly of $CaSO_4$, possesses unique engineering properties, particularly its significant expansion behavior upon exposure to water. This expansion tendency profoundly impacts the stability of tunnel openings and slope engineering. Upon contact with water, anhydrite rock undergoes hydration reactions, leading to a theoretical maximum volume increase of approximately 61%. This expansion poses a considerable challenge to slope and tunnel stability, often resulting in instability in large slopes and damage to tunnel tops and bottoms. The uneven distribution of groundwater exacerbates this challenge, potentially causing localized or widespread instability in slopes and tunnels due to uneven expansion.

In tunnels traversing anhydrite rock formations, swelling behavior can lead to significant structural challenges, including tunnel floor heave, lining destruction, and even the uplifting of entire tunnel sections and the land surface [1–3]. This swelling is primarily driven by the transformation of anhydrite ($CaSO_4$) into gypsum ($CaSO_4 \cdot 2H_2O$), a process involving indirect anhydrite dissolution and gypsum precipitation, leading to a substantial increase in the volume of sulfate minerals [4,5]. The swelling pressure, generated when volumetric swelling strain is constrained, can reach up to 60 MPa at the crystal level, 10 MPa in rock specimens, and 5 MPa at the tunnel scale [6]. Although in situ swelling pressures are generally lower than laboratory measurements, they can still challenge or exceed tunnel support pressures, significantly affecting stress and displacement distribution within the tunnel. Ignoring the swelling behavior of anhydrite rock in tunnel design can result in frequent structural damage [7,8].

After excavation, the surrounding rock deforms due to hydrostatic far-field stress and swelling behavior. To leverage the self-supporting effect of the surrounding rock, initial support structures are designed to allow a certain degree of deformation, with excavation lines planned to accommodate this deformation. Exceeding the reserved deformation value can lead to intrusion phenomena and lining structure failure. Tunnel reliability is assessed by comparing the radial displacement at the tunnel wall against the reserved deformation value using the convergence constraint method.

Recent research has developed performance functions based on tunnel analytical solutions to analyze tunnel reliability [9–13]. These solutions, based on criteria such as Hoek–Brown or Mohr–Coulomb, include elastic–plastic, elastic–brittle–plastic, and elastic strain softening models [14–17]. However, these models do not account for the impact of swelling behavior on stress and displacement. Miao et al. [18,19] introduced the humidity stress field theory for clay-sulfate rocks, providing a method to analyze stress and displacement in rocks exposed to water inflow. This theory equates the effects of humidity changes to those of temperature changes in materials, suggesting that stress fields generated by humidity changes can be analyzed in a manner similar to temperature-induced stress fields. This approach, typically employed for rocks with high water-absorbing minerals such as mudstone and shale, is expanded upon in this study to include anhydrite rocks. Anhydrite rocks exhibit swelling behavior driven by sulfate hydration over prolonged periods. The resulting analytical solution captures the ongoing influence of swelling behavior on stress and displacement distribution within anhydrite tunnels. Consequently, this paper addresses two crucial objectives: developing an elastic swelling analytical solution tailored for anhydrite rock tunnels while considering swelling behavior and assessing the reliability of such tunnels.

2. Analytical Solution for Elastic Swelling in Anhydrite Rock Tunnels

Miao et al. and Wu et al. [18–20] extensively derived the mathematical model of humidity stress field theory in the analysis of expansive rock engineering. Cheng et al. [21,22] conducted experimental research on the above theory. Chen and Wu et al. [19,21,23] further validated this theory through engineering practice. The theory of humidity stress field has been proven to be an effective theoretical method for studying the stress–strain relationship of the expansive surrounding rock. Therefore, this study draws on this theory to establish an elastic analytical model for hard gypsum rock tunnels. The validation process of the adopted theory is as follows:

Figure 1 depicts a circular tunnel excavation within a uniform, homogeneous, and isotropic anhydrite rock mass. This mass, with a radius denoted as R_0, is under hydrostatic far-field stress (P_0) and uniform support pressure (P_s). The prolonged swelling behavior of the anhydrite rock continually modifies the stress and deformation patterns of the surrounding rock over time.

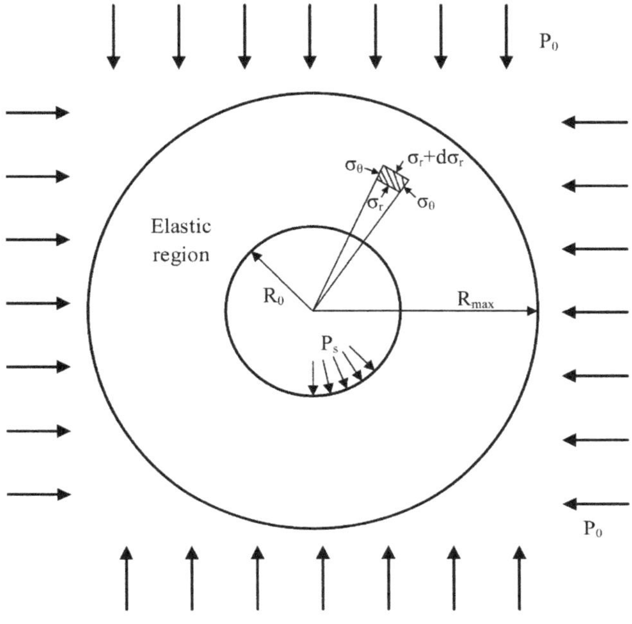

Figure 1. A circular anhydrite rock tunnel in a hydrostatic in situ stress field.

It is essential to recognize that the stress equilibrium equation remains unaffected by the root causes of stress, and likewise, the relationship between displacement and strain remains independent of causation. Therefore, in this context, both the equilibrium differential equation and the geometric equation align with those of a standard elastic problem encountered in circular axisymmetric tunnel analysis. These equations serve as the cornerstone for comprehending and forecasting the behavior of the tunnel structure under the provided conditions. Specifically, as shown in the following formula:

$$\frac{d\sigma_r}{dr} + \frac{\sigma_r - \sigma_\theta}{r} = 0 \tag{1}$$

$$\begin{cases} \varepsilon_\theta - \frac{u_r}{r} = 0 \\ \varepsilon_r - \frac{du_r}{dr} = 0 \end{cases} \tag{2}$$

where σ_r and σ_θ are the radial stress and circumferential stress, respectively, ε_r and ε_θ are the radial strain and annular strain, respectively, and u_r is the radial displacement.

This research integrates the humidity stress field theory to tackle the swelling behavior observed in anhydrite rock. This theory suggests a quasi-coupled relationship between the humidity and stress fields, following Hooke's law for the stress–strain relationship. In this analysis, it is assumed that the mechanical properties of the surrounding rock, notably the elastic modulus and Poisson's ratio, remain constant within a specified range of moisture content. While this assumption is essential for deriving analytical solutions, it is crucial to acknowledge the nonlinear effects of these factors, necessitating thorough consideration in more detailed numerical assessments.

Within the scope of this study, complexities such as the seepage movement of groundwater and its resulting influence on swelling pressure are not considered. This simplification is grounded in the assumption that anhydrite rock can completely absorb water if the initial humidity level is below the saturation threshold. Additionally, it is hypothesized that the initial humidity distribution within the surrounding rock follows a linear increase relative to the radial distance from the tunnel center.

$$W = w_0 r / R_0 \quad (3)$$

where w_0 is the initial humidity when $r = R_0$.

In the context of plane stress, the strain component in polar coordinates is as described by Miao et al. and Wu et al. [18,19]:

$$\begin{cases} \varepsilon_r = \frac{1}{E}(\sigma_r - \nu\sigma_\theta) + \alpha\Delta W \\ \varepsilon_\theta = \frac{1}{E}(\sigma_\theta - \nu\sigma_r) + \alpha\Delta W \\ \gamma_{r\theta} = \frac{2(1+\nu)}{E}\tau_{r\theta} \end{cases} \quad (4)$$

where E is the elastic modulus, ν is Poisson's ratio, α is the linear swelling coefficient, ΔW is the change in humidity, $\gamma_{r\theta}$ is the shear strain, and $\tau_{r\theta}$ is the shear stress.

The equilibrium differential equation is expressed as follows:

$$\begin{cases} \frac{\partial \sigma_r}{\partial r} + \frac{1}{r}\frac{\partial \tau_{r\theta}}{\partial \theta} + \frac{\sigma_r - \sigma_\theta}{r} - \frac{E\alpha}{1-\nu}\frac{\partial \Delta W}{\partial r} = 0 \\ \frac{1}{r}\frac{\partial \sigma_\theta}{\partial \theta} + \frac{\partial \tau_{r\theta}}{\partial r} + \frac{2\tau_{r\theta}}{r} - \frac{E\alpha}{1-\nu}\frac{1}{r}\frac{\partial \Delta W}{\partial \theta} = 0 \end{cases} \quad (5)$$

To address the plane strain problem within this context, a transformation of the material parameters in the constitutive equation is necessary. Specifically, the elastic modulus (E) is modified to $E/(1-\nu^2)$, where ν is Poisson's ratio, which in turn is adjusted to $\nu/(1-\nu)$. Additionally, the linear swelling coefficient (α) is recalibrated to $(1+\nu)\alpha$. This adjustment is crucial as the tunnel traverses through the anhydrite formation, which is closely intertwined with adjacent geological strata, leading to an axial strain that effectively approaches zero. Therefore, the constitutive equation, when expressed in polar coordinates for the plane strain scenario, is modified to account for these variations, thereby ensuring a more precise portrayal of the mechanical behavior of the anhydrite rock under tunneling conditions.

$$\begin{cases} \sigma_\mathrm{r} = \frac{(1-\nu)^2 E}{(1-2\nu)(1-\nu^2)}\left(\varepsilon_\mathrm{r} + \frac{\nu}{1-\nu}\varepsilon_\theta\right) - \frac{E\alpha\Delta W}{1-2\nu} \\ \sigma_\theta = \frac{(1-\nu)^2 E}{(1-2\nu)(1-\nu^2)}\left(\varepsilon_\theta + \frac{\nu}{1-\nu}\varepsilon_\mathrm{r}\right) - \frac{E\alpha\Delta W}{1-2\nu} \end{cases} \quad (6)$$

In employing the humidity stress field theory to simulate the swelling behavior of anhydrite rock, a prevalent assumption is that the deformation modulus during swelling aligns with the elastic modulus. However, experimental observations on anhydrite rock swelling reveal a significant discrepancy; the deformation modulus in the swelling stage is notably lower than the elastic modulus. This discrepancy is illustrated in Figure 2, where equating the deformation modulus with the elastic modulus results in an overestimation of stress levels.

In this study, the swelling behavior is likened to the effects of pore water pressure, effectively reducing the effective stress. Here, effective stress is conceptualized as the difference between the elastic stress and the swelling-induced stress. To accurately characterize the stress–strain relationship during swelling, a distinct swelling modulus, denoted as *Es*, is introduced. Simultaneously, the elastic modulus is specifically denoted as *Ee* (instead of the generic E) to differentiate it from the swelling modulus. Consequently, the constitutive equation, when formulated in polar coordinates, incorporates these adjustments to offer a more precise representation of the stress–strain dynamics during the swelling process of anhydrite rocks.

$$\begin{cases} \sigma_\mathrm{r} = \frac{(1-\nu)^2 E_\mathrm{e}}{(1-2\nu)(1-\nu^2)}\left(\varepsilon_\mathrm{r} + \frac{\nu}{1-\nu}\varepsilon_\theta\right) - \frac{E_\mathrm{s}\alpha\Delta W}{1-2\nu} \\ \sigma_\theta = \frac{(1-\nu)^2 E_\mathrm{e}}{(1-2\nu)(1-\nu^2)}\left(\varepsilon_\theta + \frac{\nu}{1-\nu}\varepsilon_\mathrm{r}\right) - \frac{E_\mathrm{s}\alpha\Delta W}{1-2\nu} \end{cases} \quad (7)$$

Consistent with the tenets of the humidity stress field theory, a dedicated swelling experiment was conducted on anhydrite rock. This experiment played a crucial role in establishing a quantitative relationship between water absorption and time, a fundamental aspect in comprehending the swelling dynamics of anhydrite rock. As outlined in the study by Chen et al. [21], this relationship is essential for understanding the temporal aspects of anhydrite rock's response to moisture exposure. The experiment's findings establish a foundational understanding of how anhydrite rock's water absorption rate evolves over time, providing essential insights into the swelling behavior under different humidity conditions. This relationship is integral to the development of more accurate models for predicting the swelling-induced changes in anhydrite rock formations.

$$A_{\mathrm{t}} = A_{\mathrm{max}}(1 - \mathrm{e}^{-at}) \tag{8}$$

where A_t is the water absorption rate at time t, A_{max} is the water absorption rate at the end of swelling, and a is the water absorption coefficient, which is used to describe the speed of water absorption.

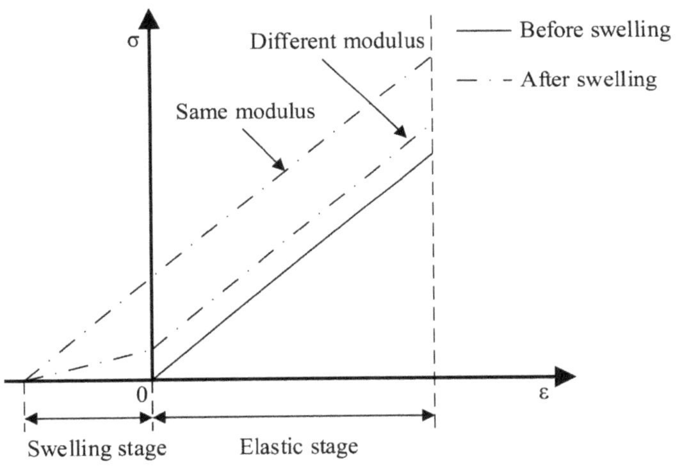

Figure 2. The elastic swelling constitutive model.

The constitutive equation, essential for analyzing the swelling behavior of anhydrite rock, incorporates the established relationship between water absorption and time. This incorporation is crucial for understanding the dynamics of swelling, especially in scenarios where the water content within a unit of surrounding rock is insufficient for the complete hydration of Ca_{SO4} in that unit. In such scenarios, the water is fully absorbed by the rock, resulting in distinct swelling characteristics. The resulting constitutive equation encapsulates the essence of elastic swelling evolution, integrating the temporal aspect. This equation stands as a fundamental tool in forecasting and comprehending the swelling behavior of anhydrite rock over time, offering a more nuanced and comprehensive perspective on the rock's response to fluctuating water content. It is expressed as follows:

$$\begin{cases} \sigma_{\mathrm{r}} = \frac{(1-\nu)^2 E_{\mathrm{e}}}{(1-2\nu)(1-\nu^2)}(\varepsilon_{\mathrm{r}} + \frac{\nu}{1-\nu}\varepsilon_{\theta}) - \frac{E_{\mathrm{s}}\alpha W(1-e^{-at})}{1-2\nu} \\ \sigma_{\theta} = \frac{(1-\nu)^2 E_{\mathrm{e}}}{(1-2\nu)(1-\nu^2)}(\varepsilon_{\theta} + \frac{\nu}{1-\nu}\varepsilon_{\mathrm{r}}) - \frac{E_{\mathrm{s}}\alpha W(1-e^{-at})}{1-2\nu} \end{cases} \tag{9}$$

In tunnel engineering, the radial stress encountered by the tunnel wall after excavation corresponds to the applied support pressure, a critical factor in comprehending tunnel stability mechanics. The outer boundary of the elastic swelling rock mass, pivotal in this context, is influenced by its neighboring rock mass and the prevailing ground stress conditions. This study explores a scenario where radial stress comprises both swelling stress and hydrostatic far-field stress, a situation commonly encountered in practical engineering. The boundary condition, crucial in this analysis, is mathematically represented by an equation that succinctly captures the relationship between these stress components. This equation plays a key role in modeling the stress dynamics around the tunnel, offering insights essential for effective tunnel design and stability assessment. The boundary condition equation can be expressed as:

$$\begin{cases} \sigma_{\mathrm{r}}(r = R_0) = P_{\mathrm{s}} \\ \sigma_{\theta}(r = R_{\max}) = P_0 + E_{\mathrm{s}}\varepsilon_{\mathrm{s}}(r = R_{\max}) \end{cases} \tag{10}$$

Drawing upon the principles of elastic mechanics, this study integrates an equilibrium differential equation, a geometric equation, and an elastic swelling evolution constitutive equation. This comprehensive approach facilitates the accurate determination of both radial and circumferential stresses in the surrounding rock. Such calculations are indispensable for comprehending stress dynamics in the context of tunnel stability and design.

$$\sigma_{\mathrm{r}} = -\frac{E_{\mathrm{e}}\alpha}{(1-\nu^2)r^2}\left(\int_{R_0}^{r}\frac{E_{\mathrm{s}}}{E_{\mathrm{e}}}W(1-e^{-at})rdr + C_1\right) + C_2 \tag{11}$$

$$\sigma_{\theta} = \frac{E_{\mathrm{e}}\alpha}{(1-\nu^2)r^2}\left(\int_{R_0}^{r}\frac{E_{\mathrm{s}}}{E_e}W(1-e^{-at})rdr + C_1 - \frac{E_{\mathrm{s}}}{E_{\mathrm{e}}}W(1-e^{-at})r^2\right) + C_2 \tag{12}$$

$$C_1 = \frac{R_{\max}^2 R_0^2(1-\nu^2)\left[\sigma_{\theta}(r = R_{\max}) - P_{\mathrm{s}} - \frac{E_e\alpha}{(1-v^2)R_{\max}^2}\left(\int_{R_0}^{R_{\max}}\frac{E_{\mathrm{s}}}{E_{\mathrm{e}}}W(1-e^{-at})rdr - \frac{E_{\mathrm{s}}}{E_e}W(1-e^{-at})R_{\max}^2\right)\right]}{E_e\alpha(R_0^2 + R_{\max}^2)} \tag{13}$$

$$C_2 = \frac{E_{\mathrm{e}}\alpha}{(1-\nu^2)R_0^2}C_1 + P_{\mathrm{s}} \tag{14}$$

where C_1 and C_2 are general solution parameters, which can be determined by boundary conditions.

Utilizing Equations (9), (11) and (12), the circumferential strain can be calculated, and the radial displacement of the surrounding rock can be derived using the geometric equation. Post-excavation displacement occurs due to stress incrementation. By employing this increment, particularly the radial and annular stresses minus the hydrostatic far-field stress, the post-excavation radial displacement can be articulated as follows:

$$u_{\mathrm{r}} = \frac{1-\nu^2}{E_{\mathrm{e}}}\left[(\sigma_{\theta} - P_0) - \frac{v}{1-v}(\sigma_{\mathrm{r}} - P_0)\right]r + \alpha(1+v)\frac{E_{\mathrm{s}}}{E_{\mathrm{e}}}W(1-e^{-at})r \tag{15}$$

3. Application

The Lirang tunnel, located in Chongqing, China, traverses an anhydrite rock formation and spans approximately 300 m (refer to Figure 3). This formation is primarily composed of 94% $CaSO_4$ and 6% other minerals. Above the anhydrite layer lies a limestone stratum averaging 120 m in thickness. The area encompasses two significant reservoirs and multiple streams. The surrounding rock is classified as Grade V, characterized by its relatively fragmented nature. The main lining of the tunnel, designed to accommodate building boundary and auxiliary structure requirements, adopts a triple-arched sidewall configuration with an arch height of 7.05 m and a radius of 5.45 m (as depicted in Figure 4).

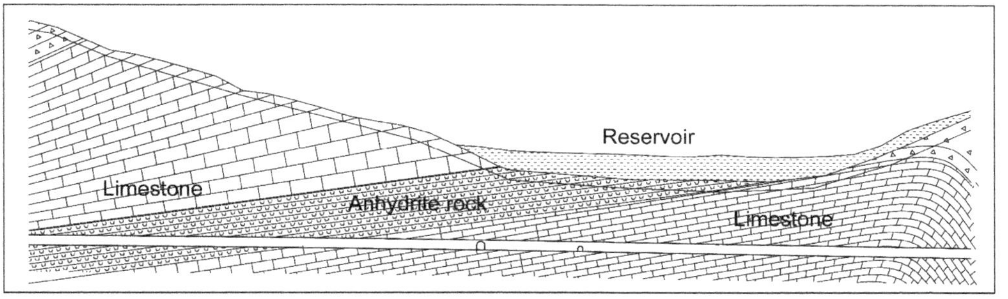

Figure 3. Geological section of the Lirang tunnel.

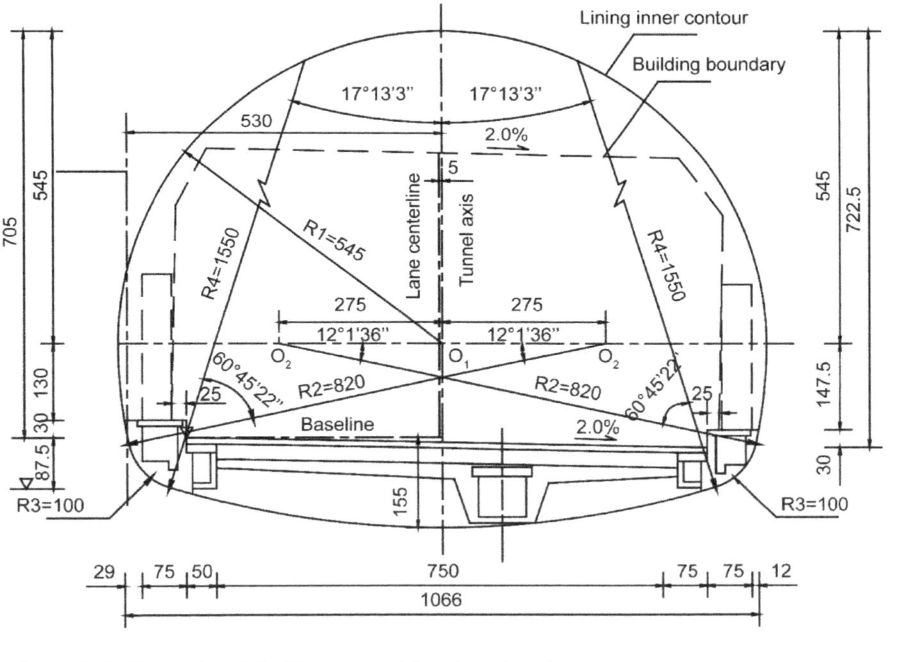

Figure 4. Outline design of the Lirang tunnel (length unit: cm).

To validate the analytical solution and evaluate the impact of anhydrite rock's swelling behavior on stress and displacement distribution, parameters were selected based on laboratory experiments and specific tunnel design criteria. The elastic modulus and Poisson's ratio were determined through uniaxial testing of the anhydrite rock. The initial humidity at $r = R_0$ represents the average natural moisture content of the anhydrite rock and was from the natural water content test. Swelling modulus, linear swelling coefficient, and water absorption coefficient were derived from swelling experiments on anhydrite rock [21,24]. The tunnel's radius, the maximum thickness of the anhydrite formation, and the hydrostatic far-field stress were determined from the tunnel design. Considering the grade and integrity of the surrounding rock, a reduction coefficient of 0.15 was applied between the rock mass's elastic modulus and the experimentally measured rock elastic modulus [25]. These parameters are detailed in Table 1.

Table 1. The values of parameters in the analytical solution.

Parameters	E_e/MPa	ν	w_0/(%)	E_s/MPa	α	a	R_0/m	R_{max}/m	P_0/MPa
Reference value	735.45	0.311	1.5	94.383	0.001	0.034	5.45	30	3

Figure 5 depicts the stress and displacement distribution under unsupported conditions. At zero swelling time, the radial stress at the tunnel wall measures 0 MPa, while the circumferential stress peaks at 5.81 MPa due to stress concentration. Subsequently, the radial stress significantly rises, gradually converging with the hydrostatic far-field stress. In contrast, the circumferential stress diminishes as the radius increases, eventually approaching the hydrostatic far-field stress. Comparable patterns are observed under varying swelling time conditions. At swelling times of 30 and 365 days, the radial stress at the tunnel wall remains at 0 MPa, while circumferential stresses measure 6.62 MPa and 7.08 MPa, respectively. This pattern suggests that swelling deformation in circumferential directions at the tunnel wall is mutually constrained, resulting in swelling stress. As swelling time advances, both radial and circumferential stresses intensify at the same location, indicating the swelling behavior of the surrounding rock. However, the stress increment decreases with increasing distance from the tunnel wall.

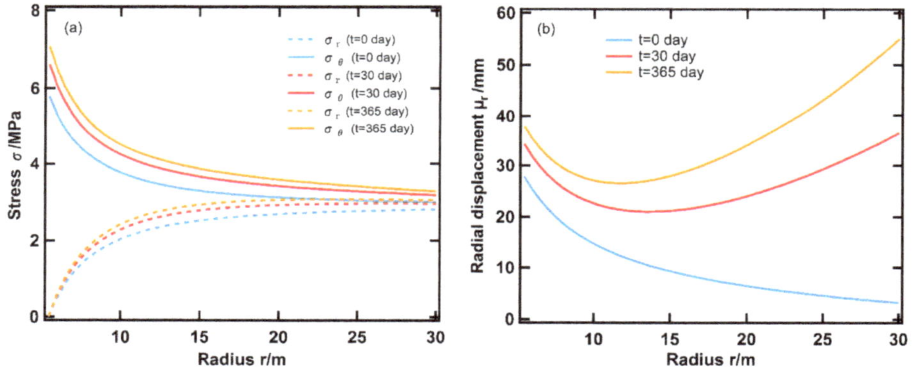

Figure 5. The stress and radial displacement distributions of different swelling times under unsupported conditions: (**a**) the stress distribution; and (**b**) the radial displacement distribution.

Additionally, as stated in the original text [20], we conducted a quantitative and qualitative comparative analysis of the data in Figure 5 through finite element simulation. This is why we have validated the results shown in Figure 5 for the stress distribution and radial displacement within the tested periods of 0, 30, and 365 days.

According to Figure 5b, radial displacement at the tunnel wall measures 27.86 mm, 34.26 mm, and 37.87 mm for swelling times of 0, 30, and 365 days, respectively. Initially, radial displacement decreases with radius, reflecting the limited influence range of tunnel excavation on the surrounding rock. However, for swelling times of 30 and 365 days, radial displacement first decreases and then increases with radius. This phenomenon is attributed to the higher initial humidity and resultant swelling deformation in the anhydrite rock at larger radii. Beyond a certain radius, swelling behavior predominantly dictates radial displacement.

Post-excavation, the displacement of tunnel walls significantly impacts tunnel design and construction. As illustrated in Figure 6, radial displacement at the tunnel wall stabilizes after 150 days of swelling, corresponding to the reduction in moisture content in the anhydrite rock and the subsequent decrease in swelling behavior. It is noteworthy that radial displacement at the tunnel wall decreases significantly with increased support pressure for the same duration of swelling.

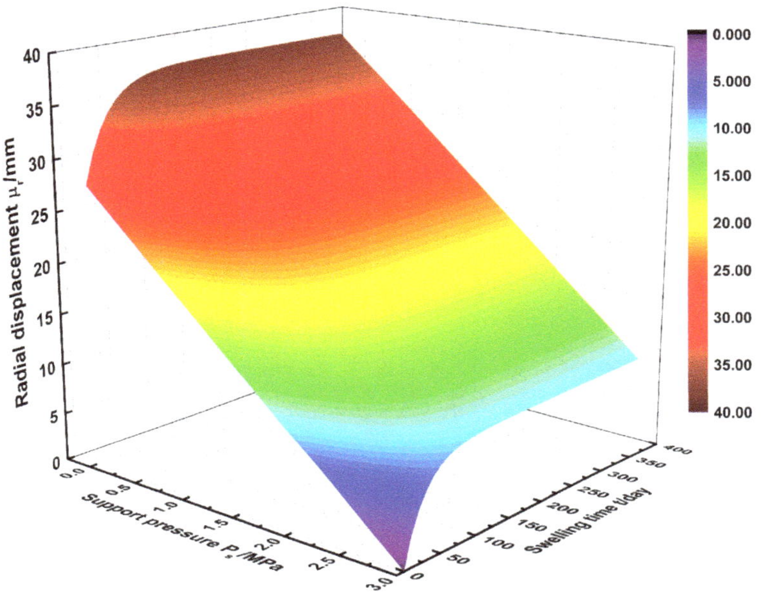

Figure 6. Synergistic effect of support pressure and swelling time on radial displacement at the tunnel wall.

4. Reliability Analysis

4.1. The Spreadsheet for the Reliability Solution

This study utilizes the spreadsheet solution method based on the First Order Reliability Method (FORM) developed by Low and Tang [26,27]. This approach is noteworthy for its straightforward input process and suitability for handling complex functions. To ascertain the reliability of the anhydrite rock tunnel, the following steps were undertaken:

(1) A performance function was formulated for reliability analysis, defined as follows:

$$g(t) = u_{cv} - u_{R_0}(t) \tag{16}$$

where u_{cv} represents the tunnel's reserved deformation. If the displacement at the tunnel wall exceeds this critical threshold, it leads to the intrusion of the lining. The magnitude of the reserved deformation is influenced by various factors, such as the quality of the rock mass and the active span of the tunnel. $u_{R0}(t)$ is the result of displacement at the free face, i.e., $r = R_0$ changing with expansion time, which is a time-dependent function. Therefore, $g(t)$ is a time-dependent functional function.

(2) This study involves categorizing variables for reliability analysis and determining their distribution types and parameters. In this context, variables such as u_{cv}, P_0, P_s, R_0, R_{max}, and t are treated as deterministic variables. The values of u_{cv} and P_s are determined within specified constraints by tunnel engineers. For the Lirang tunnel, u_{cv} and P_s are set at 50 mm and 0.5 MPa, respectively. Given the inherent uncertainties in geotechnical parameters, variables such as E_s, E_e, v, α, a, and w_0 are categorized as random variables. The types and parameters of their distributions are outlined in Figure 7 [13,28]. As shown in Figure 7, for a normally distributed random variable, para1 represents the mean, and para2 represents the variance. The solution in Column x^*(cells G7:G12) of Figure 7 denotes

the design point. The x^* values shown in cells G7:G12 render Equation (16)—performance function $g1(x)$—equal to zero.

	A	B	C	D	E	F	G	H	I	J	K	L	M	N
1				P_0/Mpa	P_s/Mpa	R_0/m	R_{max}/m	t/day	C_1	C_2				
2				3	0.5	5.45	30	0	69.14958	2.920129				
3														
4														
5			Mean value	Standard deviation						Correlation matrix				
6	variable	Distribution	para1	para2	para3	para4	x^*			R				n_i
7	E_z	Normal	94.383	6.621			94.38373	1	0	0	0	0	0	0.000111
8	E_e	Normal	735.45	50			459.3828	0	1	-0.5	0	0	0	-5.52134
9	v	Normal	0.311	0.09			0.716869	0	-0.5	1	0	0	0	4.509654
10	α	Normal	0.0011	0.00019			0.0011	0	0	0	1	0	0	-9×10^{-6}
11	a	Normal	0.034	0.007			0.034	0	0	0	0	1	0	-5.8×10^{-5}
12	w_0	Normal	1.5	0.203			1.500001	0	0	0	0	0	1	-7.04×10^{-6}
13														
14														
15				Performance function			g(t)	β	Φ(-β)					
16				$g(t)=0.05-u$ (r=R_0)			-3.3×10^{-9}	5.8791	-2.06×10^{-9}					
17														
18														
19														

Figure 7. The spreadsheet for the reliability solution.

(3) Assessing correlations among random variables is a crucial aspect of this analysis. Particularly, the variable E_e demonstrates a negative correlation with v, indicated by a correlation coefficient of -0.5 [13]. This suggests an inverse relationship between these two parameters. All other random variables in this study are regarded as independent, with no interdependencies influencing their behavior or impact.

(4) The optimization process was carried out using the constraint optimization feature integrated within Excel. This method involved leveraging Excel's built-in optimization solver to minimize the reliability index β. The optimization was constrained by the condition $g(t) = 0$, ensuring that the solution aligns with the specified performance function. During the optimization process, the initial values assigned to the variable x^* were set as their respective mean values, establishing a balanced starting point for the optimization procedure.

4.2. The Influence of Variables on Reliability

Aligned with the "Uniform Standard for Reliability Design of Building Structures", this study adopts a target reliability index of 4.2. As illustrated in Figure 8a, the reliability of the Lirang tunnel fluctuates with varying swelling durations and support pressures. Particularly noteworthy is that with a designed support pressure of 0.5 MPa, the tunnel's reliability index progressively diminishes with increasing swelling time, falling below the target threshold after 30 days. Conversely, increasing the support pressure enhances the tunnel's reliability index. A support pressure of 1.2 MPa ensures that the tunnel's reliability index remains compliant with the target requirements, even amidst swelling effects. Figure 8b depicts the tunnel's reliability concerning swelling time for different levels of reserved deformation. Achieving a reserved deformation of 59 mm enables the tunnel's reliability index to satisfy the target requirements under swelling conditions.

This analysis reveals that the initial design of the Lirang tunnel does not meet the target reliability criteria when accounting for swelling effects. However, by adjusting design variables such as support pressure and reserved deformation, the desired reliability target can be achieved. Figure 9 illustrates the combined impact of support pressure and reserved deformation on tunnel reliability over a swelling period of 150 days.

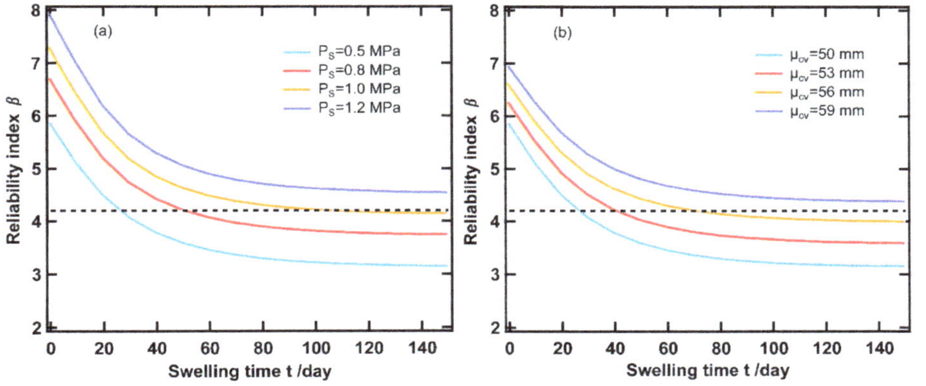

Figure 8. The reliability of the Lirang tunnel varies with swelling time under different support pressures and reserved deformation: (**a**) the reserved deformation remains at 50 mm; and (**b**) the support pressure remains at 0.5 MPa.

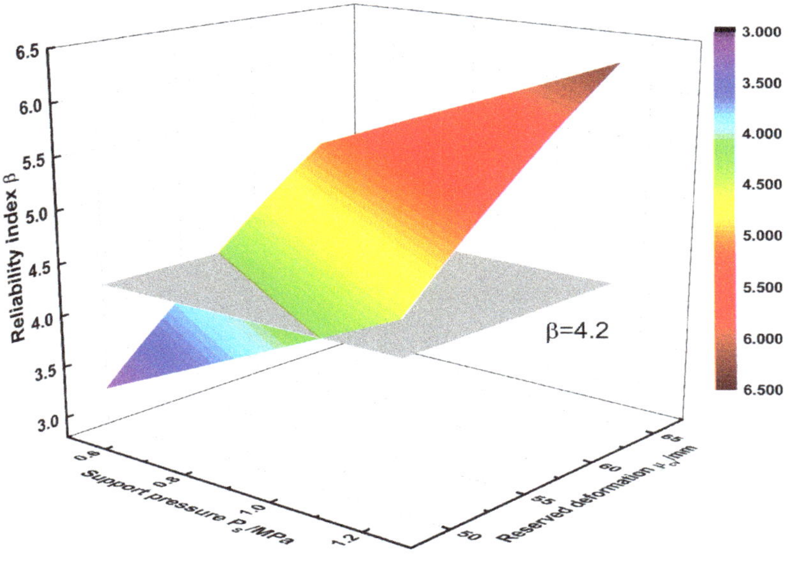

Figure 9. Synergistic effect of support pressure and reserved deformation on the reliability of the Lirang tunnel.

In summary, the reliability of the anhydrite tunnel is intricately tied to factors such as support pressure, swelling duration, and reserved deformation. Through careful adjustment and optimization of these controllable variables during the construction phase, the tunnel's reliability under swelling conditions can be enhanced to meet predefined reliability standards.

5. Discussion

5.1. Elastic Swelling Boundary Condition

When formulating the elastic swelling analytical solution, the external boundary condition for the anhydrite surrounding rock is set with the circumferential stress accounting for both the ground stress and swelling stress. However, in practical scenarios, these outer boundary conditions are closely tied to the characteristics of the adjacent rock mass.

To precisely and scientifically determine these external boundary conditions and better represent the stress and displacement distribution of the surrounding rock, the integration of an ideal elastic rock mass surrounding the elastic swelling region is proposed. This approach is depicted in Figure 10. Here, the radial stress at the outer boundary of this idealized elastic region is equated to the hydrostatic far-field stress. Simultaneously, the radial stress at the inner boundary matches that observed in the elastic swelling region. Consequently, the boundary condition equation can be expressed as follows:

$$\begin{cases} \sigma_{\mathrm{r}}(r = R_0) = P_{\mathrm{s}} \\ \sigma_{\mathrm{r}}^{\mathrm{es}}(r = R_{\max}) = \sigma_{\mathrm{r}}^{\mathrm{e}}(r = R_{\max}) \\ \sigma_{\theta}(r = R_{\mathrm{e}}) = P_0 \end{cases} \tag{17}$$

where σ_r^{es} and σ_r^{e} are the radial stresses in the elastic-swelling region and the elastic region, respectively.

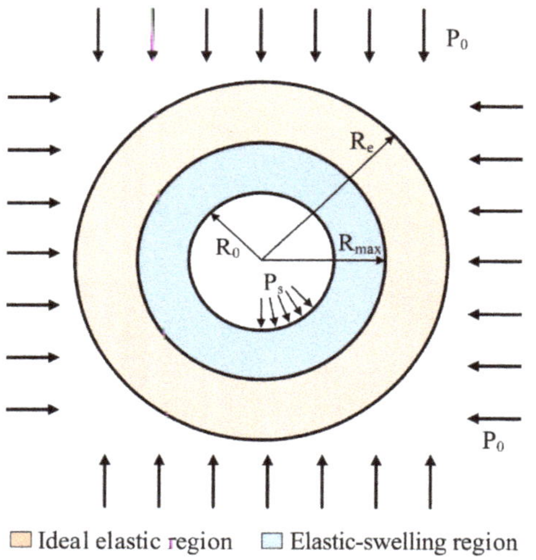

Figure 10. The circular anhydrite rock tunnel with an ideal elastic region and an elastic swelling region.

This methodology guarantees a more realistic and comprehensive comprehension of the stress distribution around the tunnel, considering the intricate interactions between the swelling anhydrite rock and its adjacent geological structures.

5.2. Application in Noncircular Tunnels

In Section 2, the analytical solution was tailored specifically for circular tunnels. However, it is crucial to acknowledge that tunnels are frequently constructed in diverse sizes and shapes, customized to their intended purpose. For example, horseshoe-shaped tunnels, as emphasized by Zhang [29], are renowned for their efficient use of underground space and their capacity to minimize soil excavation volume compared to circular tunnels.

When encountering noncircular tunnel configurations, such as semicircular or horseshoe-shaped designs (as depicted in Figure 11), a pragmatic approach is to approximate these structures as equivalent circular tunnels. This method enables the application of the circular tunnel analytical solution to a wider range of tunnel geometries, thereby expanding its utility and relevance across diverse tunneling projects. This conversion to an equivalent circular format is a pragmatic solution that streamlines the analysis and design of various tunnel shapes within the framework of the established circular tunnel model [30,31].

$$R_0 = (h + b)/4 \tag{18}$$

$$R_0 = \sqrt{h^2 + (b/2)^2}/2\cos[\tan^{-1}(2h/b)] \tag{19}$$

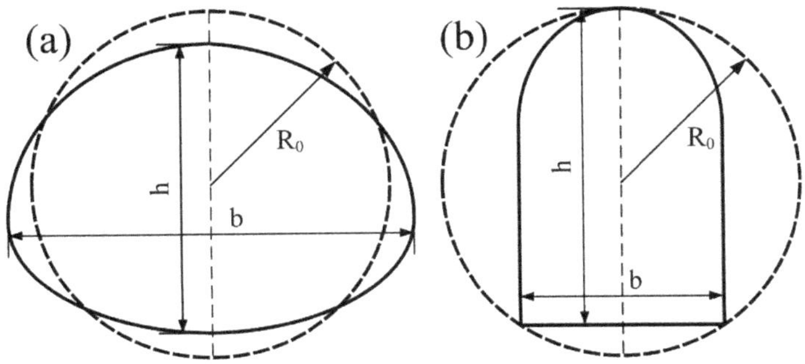

Figure 11. Conversions of noncircular tunnels into circular tunnels: (**a**) semicircular tunnel; (**b**) horseshoe-shaped tunnel.

Equations (18) and (19) represent the transformation formulas for semicircular and horseshoe-shaped tunnels, respectively. In these equations, 'h' denotes the height of the tunnel, while 'b' signifies its span. These formulas establish a mathematical framework for converting the dimensions of noncircular tunnel geometries into their circular equivalents, thereby facilitating the application of circular tunnel analysis methods to a broader spectrum of tunnel shapes. This conversion is pivotal for aligning established analytical solutions with diverse tunnel configurations, ensuring their relevance and applicability across various engineering scenarios.

6. Conclusions

This study presents an elastic swelling analytical solution for anhydrite rock tunnels rooted in the principles of humidity stress field theory. The solution's validity is affirmed through its application to the Lirang tunnel. Additionally, this research explores the effects of support pressure, swelling duration, and reserved deformation on tunnel reliability. Key findings include:

1. Initially, radial stress intensifies while circumferential stress diminishes as the radius increases, both converging towards the hydrostatic far-field stress. Radial displacement similarly decreases. With prolonged swelling, all three parameters—radial stress, circumferential stress, and radial displacement—escalate at the same radial distance, mirroring the swelling behavior of anhydrite rock. Notably, at swelling durations of 30 and 365 days, radial displacement first decreases and then surges with increasing radius. This pattern is attributed to the swelling behavior becoming the dominant factor in radial displacement beyond a certain radius, where initial humidity is higher. After 150 days, radial displacement at the tunnel wall stabilizes, correlating with the diminishing swelling behavior as the rock's internal moisture is absorbed.

2. The original design of the Lirang tunnel falls short of meeting the desired reliability standards when considering the swelling effect. However, increasing the support pressure to 1.2 MPa or enhancing the reserved deformation to 59 mm can guarantee that the tunnel meets the target reliability criteria.

3. The reliability of an anhydrite tunnel is intricately linked to factors such as support pressure, swelling time, and reserved deformation. Through careful adjustment and

optimization of these controllable variables during construction, such as modifying support pressure and reserved deformation, the tunnel's reliability under swelling conditions can be significantly enhanced to meet predetermined reliability requirements.

Author Contributions: Conceptualization, Z.H.; Methodology, Z.Z. and Z.H.; Validation, R.W. and J.W.; Formal analysis, Z.Z., Y.Z. and Z.H.; Investigation, Z.Z.; Data curation, Z.Z., R.W., Y.Z., J.W. and Z.H.; Writing—original draft, J.W.; Writing—review & editing, Y.Z. and Z.H.; Supervision, R.W. All authors have read and agreed to the published version of the manuscript.

Funding: This research was funded by the National Natural Science Foundation of China (Grant Nos. 52208336), the Guangdong Basic and Applied Basic Research Foundation (2023A1515012826), and the Science and Technology Projects in Guangzhou (2024A04J3902), the Research on Intelligent Prediction Technology of Blasting vibration of an open-pit mine (2022-2-TD-QN006), Research Project on the Management and Optimization Platform for Geotechnical Engineering Investigation and Design Based on Data Mining Technology (GSKJ2-G02-2019), and, Research on the Application of Compressed Air Energy Storage Chamber in Abandoned Mines (30-K2022-G02), which made the work presented in this paper possible.

Data Availability Statement: Data are contained within the article.

Conflicts of Interest: Authors Zihui Zan and Ran Wang were employed by the company Guangzhou Metro Design & Research Institute Co., Ltd., Author Jianxun Wu was employed by the company East China Electric Power Design Institute Co., Ltd. The remaining authors declare that the research was conducted in the absence of any commercial or financial relationships that could be construed as a potential conflict of interest.

References

1. Alonso, E.E.; Berdugo, I.R.; Ramon, A. Extreme expansive phenomena in anhydritic-gypsiferous claystone: The case of Lilla tunnel. *Géotechnique* **2013**, *63*, 584. [CrossRef]
2. Einstein, H.H. Tunnelling in difficult ground—Swelling behaviour and identification of swelling rocks. *Rock Mech. Rock Eng.* **1996**, *29*, 113–124. [CrossRef]
3. Wu, J.; Ren, S.; Ouyang, X.; Fan, J.; Deng, G.; Li, Z. Research on Swelling of Gypsiferous Rock and Its Influence on Invert Arch of Tunnel. *J. Hunan Univ. (Nat. Sci.)* **2018**, *45*, 142–149.
4. Jeschke, A.A.; Vosbeck, K.; Dreybrodt, W. Surface controlled dissolution rates of gypsum in aqueous solutions exhibit nonlinear dissolution kinetics. *Geochim. Cosmochim. Acta* **2001**, *65*, 27–34. [CrossRef]
5. Taherdangkoo, R.; Meng, T.; Amar, M.N.; Sun, Y.; Sadighi, A.; Butscher, C. Modeling Solubility of Anhydrite and Gypsum in Aqueous Solutions: Implications for Swelling of Clay-Sulfate Rocks. *Rock Mech. Rock Eng.* **2022**, *55*, 4391–4402. [CrossRef]
6. Konstantinos, S.; Anagnostou, G.; Vrakas, A. Scale effects in relation to swelling pressure in anhydritic claystones. *Geomech. Micro Macro* **2015**, *48*, 795–800.
7. Butscher, C.; Einstein, H.H.; Huggenberger, P. Effects of tunneling on groundwater flow and swelling of clay-sulfate rocks. *Water Resour. Res.* **2011**, *47*, 198–206. [CrossRef]
8. Steiner, W. Swelling rock in tunnels: Rock characterization, effect of horizontal stresses and construction procedures. *Int. J. Rock Mech. Min. Sci. Geomech. Abstr.* **1993**, *30*, 361–380. [CrossRef]
9. Hooman, G.S.; Nowak, A.S. Reliability analysis of circular tunnel with consideration of the strength limit state. *Geomech. Eng.* **2018**, *15*, 879–888.
10. Qing, L.; Chan, C.L.; Low, B.K. System reliability assessment for a rock tunnel with multiple failure modes. *Rock Mech. Rock Eng.* **2013**, *46*, 821–833.
11. Song, L.; Li, H.Z.; Chan, C.L.; Low, B.K. Reliability analysis of underground excavation in elastic-strain-softening rock mass. *Tunn. Undergr. Space Technol.* **2016**, *60*, 66–79. [CrossRef]
12. Yang, X.L.; Li, W.T. Reliability analysis of shallow tunnel with surface settlement. *Geomech. Eng.* **2017**, *12*, 313–326. [CrossRef]
13. Li, H.-Z.; Low, B.K. Reliability analysis of circular tunnel under hydrostatic stress field. *Comput. Geotech.* **2010**, *37*, 50–58. [CrossRef]
14. Wang, Y. Ground Response of Circular Tunnel in Poorly Consolidated Rock. *J. Geotech. Eng.* **1996**, *122*, 703–708. [CrossRef]
15. Carranza-Torres, C. Elasto-plastic solution of tunnel problems using the generalized form of the Hoek-Brown failure criterion. *Int. J. Rock Mech. Min. Sci.* **2004**, *41* (Suppl. S1), 629–639. [CrossRef]
16. Hou, Z.; Liu, Y.; Han, Z.; Tang, M.; Gong, X.; Su, D.; Wang, L. Experimental Study of the Bearing Characteristics of a Novel Energy-Saving and Environmentally Friendly Pile: Drilling with Prestressed Concrete Pipe Cased Piles. *Int. J. Geomech.* **2024**, *24*, 04024035. [CrossRef]
17. Alonso, E.; Alejano, L.R.; Varas, F.; Fdez-Manin, G.; Carranza-Torres, C. Ground response curves for rock masses exhibiting strain-softening behaviour. *Int. J. Numer. Anal. Methods Geomech.* **2003**, *27*, 1153–1185. [CrossRef]

18. Miao, X.X.; Yang, C.Y.; Chen, Z.D. Humidity stress field theory in swelling rock mass. *Rock Soil Mech.* **1993**, *14*, 49–55.
19. Wu, J.; Hou, Z.; Chen, L.; Du, X.; Fan, J.; Xu, X.; Sun, C. A time-dependent swelling constitutive model for reliability analysis of anhydrite tunnel. *Int. J. Mater. Prod. Technol.* **2022**, *65*, 277–298. [CrossRef]
20. Wu, J. Study on the Watering-Swelling Evolution of Anhydrite and Its Application in the Reliability of Tunnelling. Ph.D. Dissertation, Chongqing University, Chongqing, China, 2018.
21. Chen, F.; Wu, J.; Ouyang, X.; Ren, S. The swelling tests of anhydrite based on the humidity stress field theory. *Rock Soil Mech.* **2018**, *39*, 1–9.
22. Cheng, S.; Zeng, W.; Ye, Y.; Gao, R.; Li, H. Analysis Model of Geotechnical Materials Based on Relative Humidity Field and Variable Modulus Constitutive Modell. *Adv. Eng. Sci.* **2023**, *55*, 99–108.
23. Chen, X. Analysis on Deformation of Expansive Soil Tunnel and Influence on Adjacent Buildings under Humidification Condition. Ph.D. Thesis, Hefei University of Technology, Hefei, China, 2021.
24. Ren, S.; Wu, J.X.; Ou, Y.X.; Liu, R.; Wang, L. Influence of pressured water on the swelling of anhydrite rock. *Rock Soil Mech.* **2018**, *39*, 4351–4359.
25. Institute Changjiang River Scientific Research; Northeastern University. *GB/T 50218-2014*; 2014 Standard for Engineering Classification of Rock Mass. Ministry of Housing and Urban-Rural Development of the People's Republic of China; General Administration of Quality Supervision, Inspection and Quarantine of the People's Republic of China: Beijing, China, 2014.
26. Low, B.K.; Tang, W.H. Reliability analysis using object-oriented constrained optimization. *Struct. Saf.* **2004**, *26*, 69–89. [CrossRef]
27. Low, B.K.; Tang, W.H. Reliability analysis of reinforced embankments on soft ground. *Can. Geotech. J.* **1997**, *34*, 672–685. [CrossRef]
28. Xiang, L.; Wang, J.-T.; Wei, H. Reliability analysis of rock tunnel stability based on interval non-probability under multiple failure modes. *Rock Soil Mech.* **2023**, *44*, 8.
29. Zhang, J.; Feng, T.; Yang, J.; Yang, F.; Gao, Y. Upper-bound stability analysis of dual unlined horseshoe-shaped tunnels subjected to gravity. *Comput. Geotech.* **2018**, *97*, 103–110. [CrossRef]
30. Carranza-Torres, C.; Fairhurst, C. The elasto-plastic response of underground excavations in rock masses that satisfy the Hoek–Brown failure criterion. *Int. J. Rock Mech. Min. Sci.* **1999**, *36*, 777–809. [CrossRef]
31. Liu, S.-F.; Su, Y.-H.; Li, S. Calculation of safety factor of tunnel structure based on Duncan-Fama convergence curve. *Chin. J. Geotech. Eng.* **2016**, *38*, 1307–1315.